THE FEAR OF MATHS:
How to Overcome It

Sum Hope[3]

D1635853

STEVE CHINN

SOUVENIR PRESS

First published in Great Britain in 2011 by Souvenir
Press Ltd,
43 Great Russell Street, London WC1B 3PD

Portions of this book appeared in slightly different
form in *Sum Hope* (Souvenir Press) and *Dealing with
Dyscalculia* (Souvenir Press)

ISBN 9780285640511

Typeset by M Rules

Printed and bound in Great Britain

THE FEAR OF MATHS:

How to Overcome It

Contents

	Introduction	1
1	Understanding what can cause problems with learning maths	3
2	Understanding numbers	13
3	Counting	21
4	The basic facts of maths: Addition and times tables facts (and subtraction and division, too)	27
5	Add, subtract, multiply and divide	49
6	Estimation	74
7	Fractions	75
8	Probability	83
9	Decimals	85
10	Deci, centi, milli, kilo and Mega	91
11	Percentages	97
12	Proportion / ratio	106
13	Averages	113
14	Angles	116
15	Time and clocks	120
16	Algebra	125
17	Famous formulas	133
18	People having difficulty with maths	138
19	The Dyscalculia Checklist	150

Introduction

Please read this first!

This book is written to help you do maths with more confidence and success.

Maths, usually in the shape of numbers, causes anxiety and denial ("I never was any good at maths") way beyond any other school subject. Yet the need to use maths lingers long after schooling has finished. It is a part of our everyday life, for example, we have to use money, we deal with time and we meet percentages.

This little book could not attempt to re-teach all of the maths you didn't learn at school, nor should it. I think that most people know more maths than they realise, so I have tried to clarify and pull together those vague, half remembered, half understood ideas. Problems are usually rooted a long way back in the maths, so, if for example, you can't do division, you may have to revisit subtraction and place value.

I want to try and convince you that maths is based on a few ideas and concepts, all of which link together. It will be the links that help you to understand the maths and thus reduce your need to remember what must often seem like an overload of meaningless facts and procedures.

I have tackled some of the areas of maths which are most needed in life, supported by trying to give an understanding of the basic ideas of mathematics. I have attempted to give clear explanations, but you will still need to involve yourself in the learning process and persevere . . . it will all become clear in due course.

I have long recognised that not everyone uses or relates to the same methods for using mathematics. Throughout the book I have tried to explain alternate methods. No one method is better than any other, but a particular method might suit you better. The only way you will find out is to give all methods a proper try and then choose the one that works best for you. Also, trying the different methods may help you to form a better understanding of maths processes. This is because, although the methods may appear to be different, they have to be based on the same mathematical principles.

Many of the ideas in maths crop up again and again, often in different disguises. This has a minus and a plus effect. The minus effect is that if you do not understand the idea, you will probably fail to understand it each time it is used. The plus effect is that if you can get even some understanding of the idea, then each extra time it occurs you should use the new experience to strengthen your understanding. I have tried to focus on this plus!

Unfortunately maths is often taught and presented in ways that make it hard to understand. It then becomes something you have to remember, which means you will probably forget enough of it to make the bits you do remember useless

Remember that maths is a skill, just like basketball or tennis. If you don't practise the skill it will fade. But, practice is only effective if you understand what you are practising. If the ideas in this book help you, they may still need, at some stage in the future, a little top-up work (revision) especially if you are not using the maths regularly. As with many skills, learning is often most effective when taken "little and often."

I have asked Pete Jarrett, a friend whose work I admire greatly, to write Chapter 18, the chapter on Case Studies, the stories of people who have have problems with maths. It illustrates the huge importance of listening to learners and it should inspire and re-assure you that you are not alone and that problems can be tackled. It may be that this is the Chapter you should read first.

1 Understanding What Can Cause Problems with Learning Maths

If you can't do mathematics it is likely to be for some very good reasons, which probably have little to do with how clever you are. There are many factors which can get in the way of learning mathematics. Some of these are listed below. You may recognise some of these factors as relevant to you. You may be unlucky enough to be affected by all of them, but even then you may have found ways to get round some or most of the difficulties these factors create. If you haven't, this book will help you to find some ways.

You may have reached the stage where you have decided that enough is enough and that you and mathematics can live without each other. I hope to persuade you to have one more try. It is a useful skill in so many aspects of life.

When you meet a problem, a good starting point is to try and understand the causes of the problem. This often helps you understand the problem itself and should make it easier to tackle. This awareness may even help you to avoid or at least reduce the influence of the problem in the future.

So, let's look at these problem factors. . . .

Anxiety

Anxiety can really get in the way of learning.

It is an accumulation, a consequence of all the other factors and difficulties and how they have affected your attempts to succeed in mathematics, and how they have affected your attitude towards carrying on working at this subject.

Anxiety is the last difficulty to occur (because it is a

consequence of all the other problems) and the first to overcome if you are to return to using mathematics and numbers. This does not mean, however, that it doesn't occur in young children.

One of the best ways to reduce this anxiety is to find some areas of success. It is important to know that everyone can do some maths. As my colleague, Richard Ashcroft says, mathematics is a subject that builds like a wall, but it is a wall that can still stand and be strong with some gaps, some missing bricks. You do not have to be perfect in all of maths to have success. For example, an ex-student of mine still cannot give an instant answer to "What is 8×7?" but he does now have a degree . . . in maths.

The work in this book will attempt to use and build on what you know.

To be good, or even just OK at mathematics you have to practise, to gain experience, but if you are anxious about maths you will probably try to avoid doing any practice at all! For you to feel more comfortable and then, hopefully confident, I have to convince you to change your mind and try some practice.

If you do suffer from maths anxiety you are certainly not alone (see for example page 143). There have been whole books written on this subject. My guess is that people who are mathematically anxious are in the majority!

There is a (free) questionnaire on maths anxiety in adults on my website, www.stevechinn.co.uk.

DO TRY THE IDEAS IN THIS BOOK. THEY ARE DESIGNED TO HELP YOU SUCCEED AND START TO OVERCOME SOME OF THAT ANXIETY.

Learning how to do maths successfully will help reduce anxiety. Set your own targets and your own speed of working. Despite many beliefs about having to do maths quickly, there is no rush! Make both of these realistic, and then slowly increase your goals. Above all . . . BEGIN.

Long term memory

One of the most common problems in mathematics is remembering the basic facts of numeracy, in particular the times table facts (such as 6 × 7 and 4 × 9). Some people find this task virtually impossible. And since this is one of the first demands from teachers of mathematics (and expectations from parents) it can create an early sense of failure and inadequacy.

YOU DO NOT NEED TO REMEMBER ALL THE 'BASIC' FACTS.

Memory can also let you down when learning addition and subtraction facts (such as 7 + 8 and 13 – 6), but these can often be worked out, quite quickly, on your fingers. These times table and addition facts are the basic building blocks of number work, but if you cannot remember them, all is not lost, there are some ideas to help (see chapter 3). These facts can be accessed by methods which are not just about memorising each fact separately. Many of the ideas in this book try to pull together, inter-link and extend the number facts and methods, so that they become mutually supportive. You practise and learn less of the facts, but use these to access more facts.

Your memory may also let you down when you try to recall a process or method, such as how to work out percentages. I will try to make each process real by relating it to something you know to give you a good understanding and add some meaning to the maths. Understanding can support memory.

If someone asks you to recall a fact from memory, say a times table fact, and your memory is a blank, it is something like looking into a deep black pit. There seems to be no way out and if remembering the fact is your only option, then indeed, there is no way out. I will try to provide some steps to bring you out of the pit.

There is a belief among some education policy makers that doing sums 'in your head' (mental arithmetic) is good for developing maths skills. Quite simply, this is not so for many people. Mental arithmetic can overload your memory, so I have included some suggestions to reduce the possibility of this problem occurring and help you tackle this activity.

Some methods for doing arithmetic are best when written, some are better to use 'in your head.' One of the reasons for memory overload is that people try to use written methods for mental arithmetic and not all written methods transfer successfully to mental arithmetic. I will attempt to suggest which methods are better to use for each case.

Any memory decays or slips away. The brain is designed to forget as well as to remember. What holds things in your mind are frequent reminders.

Then there is the way that you remind your brain. If you can put information into the brain via different experiences, then you should have a better chance of remembering what you want to remember. The more you see, hear, say or feel, that is, putting a memory into the brain by all senses, then the more likely it will be a permanent entry in your mind.

Short term memory and working memory

Short term memory is used for remembering information, usually small quantities such as a phone number, for a short time.

Working memory is used for working with information 'in your head'. Classically used for mental arithmetic. You can check working memory by having someone say a string of digits, at one second intervals, starting with three digits. When they have finished, you have to say them in reverse order. The 'working' bit comes in as you try to hold the numbers in your mind and reverse the order. Move up to 4 digits, then 5 and 6. At some stage you will be unable to remember and reverse the digits. The biggest number of digits you succeeded with indicates how many items you can deal with in your working memory.

A weakness in either or both of these, especially working memory, is very detrimental for the ability to do maths. However, research and experience suggest that many of the problems can be circumvented.

It is a great shame that maths lessons for pupils in English schools begin with mental arithmetic. Those with weak working memories are very likely to experience failure at the start of every lesson. Failure does not motivate!

Words and language

Learning is about receiving effective communication which should then lead to understanding. Unfortunately the vocabulary and language of maths is often like a foreign language.

Sometimes the words people use when talking about arithmetic are confusing. I have a similar problem when a fluent computer expert starts explaining new software to me. It seems to me that they have a language of their own (which, of course, they do!)

One source of possible confusion in our early experience of maths is that we use more than one word for a particular mathematics meaning, for example, to talk about adding we can say, for example, . . . 6 more than 3, 17 and 26, 52 plus 39, 15 add 8. Having more than one way to express an idea or concept can challenge our need for consistency.

Another potential problem is that sometimes the words we use have other, non-mathematical meanings, for example 30 take away 12, 18 shared between 3 people.

Sometimes the same words can mean two things, for example; 'What is 5 *more* than 8?' is addition, but 'Emily has 16 sweets. She has 6 *more* than Sarah. How many sweets does Sarah have?' This is subtraction.

Then there are words for the more common everyday examples of mathematics ideas that do not fit the normal pattern of other maths vocabulary around those ideas. For example with fractions we have special names for the three most common values, $\frac{1}{2}$, $\frac{1}{3}$ and $\frac{1}{4}$. These we call one half, one third and one quarter rather than one twoth, one threeth and one fourth which would fit better into the subsequent fraction pattern. The other fractions use words that are better related to the numbers in the fractions, such as one seventh ($\frac{1}{7}$) and one twentieth ($\frac{1}{20}$).

The teen numbers are another case of a break with the later pattern causing possible confusion. This is especially true for young children as they try to make sense of our number system. The way we say the teen numbers is backwards to the way we say numbers in the twenties, thirties, forties and other two figure numbers. So we say eighteen (eight ten) and write 18, then we say twenty eight, thirty eight and so on which have the figures in the correct word order 28, 38 and so on. This particular situation is exacerbated by the words eleven (which could be 'one ten and one') and twelve (which could be 'one ten and two'). Even the 'backward' words such as thirteen would be better if they were threeten (and didn't sound a lot like thirty). I suppose a small benefit of eleven and twelve is that it saves two more teenage years.

Although these examples may sound insignificant, they can be enough to start an impression in the learner that maths is confusing and inconsistent. Early maths is frequently inconsistent which does not help learners who are looking for patterns.

Also early experiences, the first experiences of trying to learn something new are very dominant in our brains. If we learn incorrectly, it is hard to subsequently unlearn.

School maths and some adult maths courses use word problems. Largely speaking, these are rarely nothing what-so-ever to do with real life and are often worded in a strange, almost non-English language, style. Singapore has the highly effective 'Singapore Model Method' for addressing this issue.

Sequences and patterns

If you can remember and recognise sequences and patterns it will help your memory. For example it is much easier to remember the seven numbers 1234567 than a random set such as 5274318.

Sequences and patterns such as 2, 4, 6, 8, 10, 12 . . . or 10, 20, 30, 40, 50, 60 . . . are very much a part of maths. You need to be able to remember them, understand them and often adapt them. So 10, 20, 30, 40 . . . can be adapted to 13, 23, 33, 43, 53. . . . Some people find this adaptation difficult. This particular example is about understanding place value.

This book will show you lots of suggestions and ideas for organising and accessing facts, often by using patterns.

IF YOU CAN'T RECALL A MATHS FACT, YOU CAN USUALLY WORK IT OUT.

I like this aspect of maths. You can use one fact to access other facts. Maths facts usually interlink. Not true for many other subjects. For example, if you know that the capital of France is Paris, you cannot use that information to find out the capital of Greece.

Speed

Not only do people expect you to do maths correctly, they often expect you to do it quickly. Both these expectations create anxiety in many people. Trying to work more quickly than you normally do will increase anxiety in you and will almost certainly make you less successful. This is like taking up jogging. You cannot convert yourself from a couch potato to a 5 minute mile runner overnight (and you may never ever reach that 5 minute goal, nor want too!). If you do want to start jogging, or squash, or oil painting or fishing you will need to learn new skills and practise them. As you practise, *providing you succeed*, you get faster and the task gets easier, or to be precise, because the task stays the same, you find the task easier. But often you have to ask about maths, is there really any need to hurry that much?

You can develop quicker ways with maths, but on the whole this is very much a secondary goal, unless you want to appear on Countdown.

Thinking Style

It seems obvious to say that not everyone thinks the same way. We have our own thinking style for all the different things we do in life. This includes having our own thinking style for maths.

It helps us learn if we understand the way we think. In fact there is a word for knowing how we think . . . meta-cognition.

Our maths thinking style is the way we work out maths problems. It is possible to simplify this individuality down a little and imagine that our own thinking style lies somewhere along a line or a spectrum. At one end of this spectrum are the inchworms and at the other end are the grasshoppers. It is usual for people to make use of both styles.

An inchworm likes to work with formulas and fixed methods.

Inchworms work step by step, preferring to write things down. They see the details of a problem.

They also tend to see numbers exactly as they are written, a sort of numerical equivalent of a literal interpretation. This can be a disadvantage if you have a limited number of facts (and procedures) ready for retrieval from memory.

Estimation is not an inchworm skill

A grasshopper often goes straight to an answer. Grasshoppers rarely write down working. They like to see the whole picture – they overview. They are intuitive, have a strong sense of number and can be confused by formulas (and see no reason to use them). They tend to see a broad value in numbers, inter-relating them to comfortable values, for example 98 is seen as a 'bit less than' 100 or 25 is seen as half and then half again of 100. Grasshoppers are good at estimation . . . a great life skill.

If someone tries to explain a grasshopper method to an inchworm, the inchworm learner will probably not relate to the method. And vice-versa. This is a fairly important situation for teachers and learners to understand.

For example, an inchworm will mentally add 340 and 98 step by step, just as if it was being done on paper, starting with the units. So 0 add 8, then 4 add 9 and finally 3 add (the carried) 1 to give an answer of 438. Inchworm workers usually like to use pen and paper to write down their method, probably as

$$3^140$$
$$+ 98$$
$$\overline{438}$$

This is a difficult procedure if you have a poor working memory.

Faced with the same question, a grasshopper will look at the 98 and round it up to 100, add 340 and 100 and subtract the 2 (which made 98 into 100), getting an answer of 438 without writing anything down.

It is best if you can learn how to make use of both thinking styles. Generally speaking, grasshoppers are better at mental arithmetic and estimating, while inchworms are good at using formulas and detailed work. So you can see that to be versatile in your maths skills you need to be able to draw on both styles of thinking.

Some people are set at the extremes of the thinking style spectrum and find it very hard to adjust to the other style.

Throughout this book you will see that some methods are more inchworm friendly and some are more grasshopper friendly. It may well be the consequence of your thinking style that makes some methods easier to understand than others.

Remember, both thinking styles have strengths and weaknesses. Ideally, you need to learn to make the best use of both.

Attitude

This is closely linked to anxiety and is a good final topic for this chapter.

One of the attitudes adopted by people who are not succeeding in maths is the attitude of not caring, not trying. This is usually based on an idea of protecting yourself from being wrong, from failing (any sensible person tries to avoid failure). So if you do not try to answer a question you cannot get it wrong. But this also means that by not allowing yourself to be wrong you are not allowing yourself to learn. I hope to encourage you to take the risk of sometimes being wrong. Even the best mathematicians make mistakes.

One of the key factors for success is a willingness to take a risk. If you look at every number problem and think "I can't even begin that" then you will not learn. You need to take that risk and experience new ideas.

Often in schools children are placed in situations where they meet a question to which they do not know the answer. Rather than be wrong, they do not try to work it out. They are withdrawing from a learning opportunity . . . understandably. My recent research suggests that this becomes a significant issue, for too many children, at age 7 years.

So . . .

You have to be involved in the learning. Learning is not a 'sit back and hope for something to happen' activity. This book will help you learn some new ways to do maths, but only if you try and then, when you feel you understand, practise the ideas. I wish I could provide a magic learning pill which you could take each day to give you instant knowledge and understanding, but I can't. The magic only comes from a combination of (hopefully) good explanations by the teacher/tutor/book and a willingness by the learner to take risks and practise.

2 Understanding Nu

When we write numbers they are made up by different ~
tions of just ten symbols / digits.

The ten symbols / digits are;

1, 2, 3, 4, 5, 6, 7, 8, 9, and 0.

All our numbers are based on groups of ten, ten units, ten tens (hundred) , ten hundreds (thousand), ten thousands and so on. It is no coincidence that we have ten fingers and that we often use them to help us count.

One of the problems around how we write numbers is that the task can seem simple, when it is actually not at all simple. It can be a case of familiarity breeds contempt. Being able to chant the numbers in order does not guarantee that the underlying number system is understood.

Numbers are such a part of our growing-up experiences that their familiarity may prevent us from really understanding them from a maths conceptual perspective. It may be a case of giving an illusion of learning.

To explain how written numbers are built up and the structure of what is known as 'place value' I have used an example which shows how you count to, and write, one hundred and eleven as 111. The three digits represent very different values, even though they represent the same value when written separately. The value depends on the place they hold, that is where they are in the order, in the full number 111.

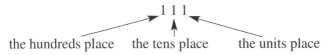

the hundreds place   the tens place   the units place

have a job as a sheep counter and that you are count-
e number of sheep (and still staying awake). You can
to the first ten sheep by using your fingers up to 9 sheep.
you get to ten, you call in a second sheep counter to record
you have one ten already counted. You *nudge* the second sheep
unter to ask him to use one finger to represent the one lot of ten
you have just counted. If you were writing the numbers, then this
is bringing in a second digit/symbol. Let's say you have counted
sixteen sheep, then you can use two of the number symbols / digits
to write this down as . . .

1 6

↑ ↑

(for the first ten you counted) | | (for the next 6 you counted)

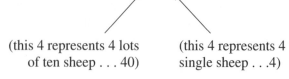

The *nudge* means that you have crossed from fingers that represent
ones to fingers that represent tens. You have crossed from units to
tens.

We use the ten digit symbols written together (in an order which
also tells us about the value of the number) to represent any number
of sheep. For example forty four sheep is written as . . .

4 4

(this 4 represents 4 lots   (this 4 represents 4
of ten sheep . . . 40)   single sheep . . .4)

So the first 4 you write represents 4 tens (40) and the second 4 represents 4 units (4)

Imagine now that you have counted ninety-nine sheep. Your second sheep counter has nine fingers counted and you also have nine fingers counted. This writes up as

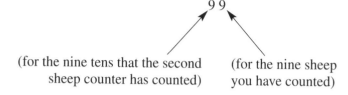

9 9

(for the nine tens that the second sheep counter has counted)   (for the nine sheep you have counted)

The first 9 you write represents 9 tens (90) and the second 9 represents 9 units (9).

Now, think what happens to the counting when the next sheep after number 99 appears. You reach another ten on your (unit) fingers. The second sheep counter also reaches ten (but his are ten lots of ten). You now need to bring in a third sheep counter to record hundreds. The second sheep counter *nudges* the third sheep counter who raises one finger which represents 1 hundred, or ten lots of ten.

This number, one hundred, is written as 100. Now we are using three symbols, three digits 1 and 0 and 0. The third sheep counter's fingers count the hundreds, which are ten lots of ten. The other two sheep counters are not showing any fingers, since there are no single sheep and no tens of sheep, zero units and zero tens.

The *nudge* means that you have crossed from fingers representing tens to fingers that represents hundreds. You have crossed from tens to hundreds.

The next sheep takes us to 101, so the second sheep counter still has zero tens to register. However, you need him there or you might think there are only 11 sheep.

Another ten sheep take us to one hundred and eleven sheep 111. You have one finger counted to represent 1 sheep. The second sheep counter has one finger counted to represent 1 lot of ten sheep and the third sheep counter has one finger counted to represent 1 lot of a hundred sheep.

So in 111, each 1 represents a different value.

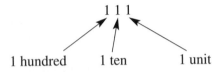

You have used three digits to represent the number one hundred and eleven. In this particular number all three digits are 1, but each 1 represents a different amount . . . 1 hundred, 1 ten and 1 unit.

This is the way all numbers are built up and written. So, when you get to thousands you bring in a fourth number symbol / digit, so, for example five thousand, six hundred and eighty nine is

5 689

Note that this thousands digit is written with a gap between the thousands digit and the hundreds digit. This is a convention designed to help you visually organise the digits in a large number. Later, with millions, another gap is introduced between the millions digit and the hundred thousands digit.

The next group uses five digit symbols, for example; twenty three thousand, four hundred and sixty seven is

23 467

then six digit numbers, for example, nine hundred and fifty four thousand, seven hundred and eighty one is

954 781

and on to a million, a seven digit number, for example three million, two hundred and fifty six thousand, nine hundred and forty two is

3 256 942

Mathematicians talk about the place value of digits. What they mean is that the value of a particular digit in a number depends on its place (or position) within that number. You have to look at the relative positions of the digits.

So, in 3 256 942 there are two 2's. The 2 between the 3 and the 5 represents 200 000, two hundred thousand and the 2 at the right hand end represents just 2, two units. So the place value of the same digit can be very different!

The most common problem with writing a number in digits from a number in words occurs when there are zeros in the number. In the million number example above there was a digit for every place value. Look back at page 15 and the number one hundred and one. There are only two number words but you need three digits to write the number, 101. The zero is needed to show that the number is more than 100. Without the zero the number is 11.

Try a slightly harder example . . . four thousand and sixty five . . . 4065.

There are three word numbers (four, sixty and five), but it is a four digit number. The zero is used because there are zero (no) hundreds. Without the zero, the number becomes four hundred and sixty five . . . 465.

Go harder again . . . five hundred thousand and six . . . 500 006. There are two word numbers (five and six) for a six digit number. Without the zeros this big number becomes 56!

How to write word numbers in digits (useful for cheques and amounts of money).

Even though cheques are being phased out as a way of transferring money, businesses often have special cheques for big sums of money (where mistakes can be expensive).

This gives a good strategy for writing numbers as digits. Work to a place value grid

million	hundred thousand	ten thousand	thousand	hundred	ten	unit*

Take the number and read it, for example:

Seven million, four thousand and ten (just in case you win the Lottery)

It says seven million so put 7 in the million slot.
Or write 7 000 000 (in pencil):

million	hundred thousand	ten thousand	thousand	hundred	ten	unit
7						

It says four thousand so put 4 in the thousand slot.
Or write the 4000 on top of the 7 000 000   7 004 000

million	hundred thousand	ten thousand	thousand	hundred	ten	unit
7			4			

It says ten so put 1 in the ten slot.
Or write in 10 on top of the 7 004 000   7 004 010

million	hundred thousand	ten thousand	thousand	hundred	ten	unit
7			4			1

Now fill in the other slots with zeros, since the word number did not contain any other digits.

million	hundred thousand	ten thousand	thousand	hundred	ten	unit
7	0	0	4	0	1	0

The word number translates to a seven digit number 7 004 010

Try a smaller number, thirty thousand and twenty four

Read the number and start to fill in a place value grid.

Thirty thousand . . .

million	hundred thousand	ten thousand	thousand	hundred	ten	unit
		3				

Or write 30 000

and twenty

million	hundred thousand	ten thousand	thousand	hundred	ten	unit
		3			2	

Or write the 20 on top 30 020

Now the four

million	hundred thousand	ten thousand	thousand	hundred	ten	unit
		3			2	4

Or write the 4 on top 30 024

Now fill in the zeros

million	hundred thousand	ten thousand	thousand	hundred	ten	unit
		3	0	0	2	4

The word number thirty thousand and twenty four writes out in figures as 30 024

A way of checking your answer

There is a basic, but only partial, check you can make on whether or not you have written a number correctly. You count the number of digits in the number.

Sometimes you read in newspapers of people receiving a five figure (they mean digit) or six digit (or figure) salary. The number of digits in a number gives some idea of the value of the number. So, the first (and lowest) five digit (figure) number is 10 000 and the last (and biggest) is 99 999 . . . quite a big range!

The first six digit (figure) number is 100 000 and the last is 999 999 (one short of a million).

- Hundreds are three digit (figure) numbers
- Thousands are four digit (figure) numbers
- Ten thousands are five digit (figure) numbers
- Hundred thousands are six digit (figure) numbers
- Millions are seven digit (figure) numbers
- Billions are ten digit (figure) numbers
- Trillions are thirteen digit (figure) numbers

It is so very important to understand how place value works. Knowing place value means that you can read numbers and know what they represent, what they are worth.

3 Counting

Most children can learn to count. This is good, because maths
begins with counting. However, their ability to count does not necessarily mean they are building the basic concepts needed for
learning maths. This early counting is, of course, in ones, one
number at a time, and is usually forwards.

If maths skills and understanding are to evolve then adults and
children need to understand the number system we use, particularly
'place value'. They also need to know how to go beyond counting
in ones, which infers seeing, appraising and using numbers in
quantities that are more than one.

Counting in ones

One of the key principles of learning maths is taking what you can
do and extending it. One of the great benefits in maths is that you
can often take what you know and extend it to things you don't
know. For example if know that $10 + 6 = 16$, you can work out that
$9 + 6 = 15$. So maths can be about adding in new learning, learning
that is building on existing understandings and knowledge. The
fact that much of maths builds and develops is both a positive and
a negative. The positive is that you have the opportunity to take an
idea or some facts and learn new ideas and new facts. The negative
is that without that understanding and development, new ideas and
facts appear in isolation with nothing to help you to remember
them.

*Just because you can do something does not mean that you
understand it. Read on. . . .*

Counting is something that most children, and adults, can do.
However, there is a difference between saying the words, one, two,

three, four, five, six, seven, eight, nine, ten, eleven, twelve, thirteen and so on and matching each word to a quantity. This matching could mean moving one item for each number, known as 'one-to-one correspondence'. Without that matching, the numbers are just words. They are not understood as representing quantities of things.

Also it is about knowing what the 1 and what the 3 represents in '13' and why thirteen is not written as 31.

Counting is the fundamental skill for arithmetic. It involves place value, maths vocabulary and sets the foundations for addition and multiplication. Counting backwards (not always as easy a transition from counting forwards as might be assumed) sets the foundations for subtraction and thus for division.

Counting, forwards and backwards, starts with counting in ones. Counting up in ones is adding one each time. Consequently counting backwards in ones is subtracting one each time. As soon as you reach ten, from counting up or from counting back, then place value becomes important. This step is sometimes known as 'crossing the tens'.

Crossing the tens is equivalent to the first sheep counter reaching ten then nudging the second sheep counter who displays one finger, which represents one ten.

The equivalent with money would be to count up in ones using one pence coins. When you reach ten 1p coins you exchange them for one 10p coin. This is the same in principle as the sheep counter's nudge.

If you were counting back, say from 15p, represented by one 10p coin and five 1p coins, then when you reached ten, represented by one 10p coin then that 10p coin would have to be traded or exchanged for ten 1p coins. The 'nudge' is now in the opposite direction.

So, if you can count up, then you are well on the way to being able to do addition problems. If you can count back then you are well on the way to being able to do subtraction problems.

Subitising

This is a word that has come to be associated with dyscalculia. Prof Brian Butterworth, the UK's foremost expert and researcher into dyscalculia, considers that the skill of subitising is a pre-requisite for acquiring mathematics skills.

Subitising is the ability to look at a random scatter of dots and know how many dots are there. For example, look at the dots below and see how quickly you can tell how many are there. Did you count them one by one? Did you cluster them? Did you 'just know'?

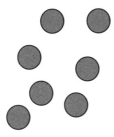

Most adults can be pretty accurate up to around 7 dots. I have worked with teenagers who, when shown three fingers, have to count them to know that there are three.

Subitising is about having a sense of quantity related to a number, knowing that the dots above can be represented by 7.

Counting and patterns

Counting in ones is the beginning. Beginning to get a 'feel' for numbers you meet as you count in ones, what they are worth and how those numbers inter-relate to each other is another step.

It may well help if some numbers are seen as visual patterns. The inter-relationships build a better sense of number, but they also start to build an understanding of addition, subtraction, multiplication and division. So, even though we are still down with those early numbers we can really get an initial understanding of arithmetic and mathematics.

The patterns are based on 1, 2 and 5:
Key links are written below each pattern

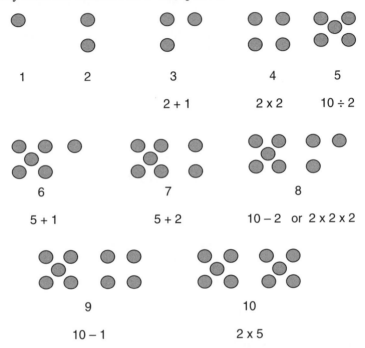

When trying to understand processes in maths, such as addition, subtraction, multiplication and division (known as the four operations), it helps if the numbers used in the explanations are not a barrier. That is, that the numbers used are clear and non-threatening in the learner's mind. In the diagrams above, although all the numbers from 1 to 10 are represented, the key numbers are:

1   2   5   and   10

These numbers are the values used in our money system, largely because they are the values that are the most useful and flexible.

So, looking at patterns, it is possible to see relationships which illustrate the four operations (add, subtract, multiply and divide) and the links between the numbers. All of the examples below can be illustrated with coins or counters set out in patterns.

For example,
(Beware of the vocabulary we use in maths. Make sure you understand it. There is more than one word used to express each operation. Hopefully this does not confuse you. Hopefully the range of vocabulary can help to clarify the concept).

For addition:

3 is 2 more than 1	3 = 2 + 1	1 plus 2 is 3	1 + 2 = 3
5 add 1 makes 6	5 + 1 = 6	1 plus 5 equals 6	3 + 3 = 6
9 and 1 make 10	9 + 1 = 10	the total of 1 and 9 is 10	1 + 9 = 10
	10 is 1 more than 9	10 = 9 + 1	

For multiplication:

2 is twice 1   2 is double 1   2 is 2 lots of 1   2 is 2 times 1 $2 = 2 \times 1$
4 is twice 2   4 is double 2   4 is 2 lots of 2   4 is 2 times 2 $4 = 2 \times 2$
    And (as often used in the 'times tables')   'Two twos are 4'
10 is twice 5 10 is double 2 10 is 2 lots of 5 10 is 2 times 5 $10 = 2 \times 5$

For subtraction:

10 take away 1 is 9 $10 - 1 = 9$ 9 is 1 less than 10 $9 = 10 - 1$
5 subtract 1 is 4   and   5 minus 1 is 4 $5 - 1 = 4$
4 is 5 minus 1 $4 = 5 - 1$

For division:

10 divided by 2 is 5 $10 \div 2 = 5$
5 is 10 divided by 2 $5 = 10 \div 2$
How many fives in 10?
What is 10 shared into 2 equal groups?
5 is half of 10 (the opposite statement is '10 is twice 5')

There is a lot of maths and maths vocabulary in these four illustrations. Some of it could be used to set the roots for later topics. Just how much is drawn out from these recent examples, including the dot/coin patterns, depends on what any individual learner wishes to learn. By going back to these basics, later concepts can be re-explored. The roots of a lot of arithmetic and much of maths are here.

Counting in twos and tens

The next step is to move on from counting on in ones to counting on in twos, fives, tens and hundreds (or any other number, but these are easier and are more important). This is close to the skill needed for addition 'sums'. When doing addition you add in 'chunks', such as twenty, rather than in ones.

It follows that counting back in twos, fives, tens and hundreds is close to the skill of subtraction.

Adding and subtracting problems are about using numbers that are more than one.

Some examples:

Counting in twos:

The even numbers
2, 4, 6, 8, 10, 12, 14, 16, 18, 20, 22, 24, 26 . . .

Note how the unit digits follow a repeating pattern, as they do with the odd numbers:

The odd numbers
1, 3, 5, 7, 9, 11, 13, 15, 17, 19, 21, 23, 25 . . .

Again there is a repeating pattern.

Counting in tens:

First example is when the unit digit is zero
10, 20, 30, 40, 50, 60, 70, 80, 90, 100 . . .

Second example is when the unit digit is a value that is not zero, for example, 7
17, 27, 37, 47, 57, 67, 77, 87, 97, 107 . . .

Again, both sequences have a pattern to help you to remember and reproduce the sequence.

4   The Basic Facts of Maths: Addition and Times Table Facts (subtraction and division, too)

Many people, including those who construct maths curricula, believe that anyone and everyone can commit these facts to memory. Sadly this is not so. However, this is not the disaster it may seem to be.

The word 'basic' suggests that these are essential facts, that they should be learned. It would be good if everyone could commit these facts to memory, but, since everyone cannot, what are the alternatives?

In the USA the words 'basic facts' have been changed to 'number combinations' to recognise that everyone cannot commit them to memory, but use strategies for accessing them instead.

If you can learn the basic facts such as $6 + 7 = 13$, $13 - 6 = 7$, $5 \times 6 = 30$, $30 \div 5 = 6$ and recall them quickly it will obviously be a big help when working with numbers. If you can't learn them then I have given, in this chapter, some suggestions you can try to help you to work them out by using strategies (that is, links to other facts). If that still doesn't work then many of the methods described later in this book to do number work go round the problem.

There is enough research to tell us that the use of strategies is one of the most effective ways of learning maths.

The ability to rote-learn the basic facts may actually give the wrong impression of success. Knowing the facts does not necessarily make you good at maths.

It is good, particularly for working memory, if you can get to these basic number facts fairly quickly, either by direct memory or by efficient strategies. You have to decide what is best for you. The time you spend on this task must be rewarding in terms of success or you will simply lose interest in the work.

With any topic in this book, never be afraid to move on for a while and then return back for another try. Sometimes knowing something about the work ahead helps you understand the relevance of the earlier work.

Addition and Subtraction Facts

These are the basic facts from $0 + 0$ to $10 + 10$ and $20 - 10$ to $0 - 0$.

They are called basic because they are the base of all additions and subtractions . . . however. . . .

Being BASIC facts means that people are expected to retrieve them in one go, from long-term memory, not by strategies. However, efficient strategies are good!

The facts are there to take you on from the inefficient process of counting in ones. And a single fact can do many things. So, for example, if you know that $3 + 8 = 11$, then you will see this used, again and again, in any example involving 8 added to 3, or 3 added to 8, such as

$$3 + 8 = 11 \qquad 8 + 3 = 11$$
$$13 + 8 = 21 \quad 3 + 18 = 21 \qquad 8 + 13 = 21 \quad 18 + 3 = 21$$
$$23 + 18 = 41 \quad 13 + 28 = 41 \qquad 18 + 23 = 41 \quad 18 + 13 = 41$$

$$30 + 80 = 110 \quad 300 + 800 = 1100 \quad 80 + 30 = 110 \quad 800 + 300 = 1100$$

This consistent pattern, using different place values, is very useful. What it means, in effect, is that in maths you can develop your knowledge of one fact into many other facts.

One of the ways you can help yourself understand maths is to use objects that you are familiar with to illustrate an idea. Coins are a good example.

A reminder: Numbers use place value (Chapter 2). A digit can be used as a unit or a ten or a hundred and so on. For example, 2 is 2 units. The 2 in 20 is 2 tens and the 2 in 200 is 2 hundreds. Money can give good visual support. In these examples, one pence coins are used as units, ten pence coins are used as tens and one pound coins represent hundreds.

1p coin: unit   10p coin: ten   £1 coin: hundred

Now try the examples 3 + 8 and 13 + 8 with money.

Set up the basic fact first. Take 3 one pence coins and add them to 8 one pence coins. Count them to show 11 one pence coins.

You could exchange (or trade) 10 of the one pence coins for 1 ten pence coin.

Note that the 'trade' is the same as the sheep counters 'nudge' (p 14)

Showing 8 + 3 with coins;

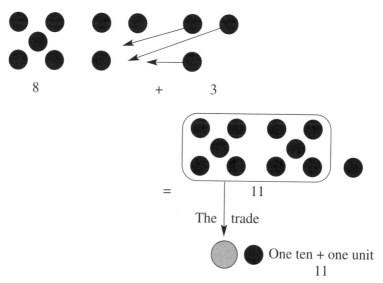

So adding 8 to 3 gives 1 ten and 1 unit, 11, eleven.

Now try 13 + 8 with coins.

Put together the coins. This gives you 1 ten pence coin and 11 one pence coins. Trade 10 of the one pence coins for 1 ten pence coins.

You now have 2 ten pence coins and 1 one pence coin, which is 21 pence. This is the same sum as the basic 8 + 3 fact, except that you started with an extra ten pence coin.

Try 3 + 18 with coins.

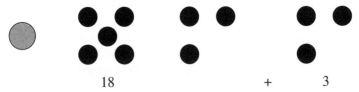

18   +   3

Now try 30 + 80 with ten pence coins.

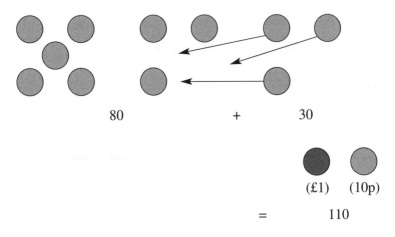

80   +   30

(£1)   (10p)

=   110

This time you can trade 10 ten pence coins for one pound. One pound is 10 × 10 p, so it can be exchanged or traded for10 lots of 10 pence. (10 × 10p is 100 pence).

In each example you were following the same procedure and using the same basic fact. This is one of the strengths of maths. You can always build on what you know to extend what you know.

Other examples of patterns;

Adding 9: Look at the patterns, that is, the consistency in the unit digits and the progression in the tens digits.

$$9 + 6 = 15$$
$$19 + 6 = 25$$
$$19 + 16 = 35$$
$$29 + 16 = 45$$
$$29 + 26 = 55$$

Subtracting 9: Again look at the patterns, consistencies and relationships in both the unit digits and the tens digits

$$15 - 9 = 6$$
$$25 - 9 = 16$$
$$35 - 9 = 26$$
$$45 - 9 = 36$$

Back to the basic facts

There are two ways to access the basic facts of addition and multiplication (times tables). Of course, these in turn lead to the subtraction and division basic facts.

You can memorise them, for example by chanting them, possibly to music. This is called rote learning. Unfortunately, it does not work for everyone. The trouble is, we don't know why, so some of those people who can rote learn find it hard to believe that there are others who cannot.

However, you can learn some efficient strategies which use the facts you do know to work out the facts you do not know. Learning to use these strategies has a bonus advantage, it will also teach you some fundamental maths processes and concepts.

A lot of this work on strategies is covered in great detail in my books "What to do when you can't learn the Times Tables" and "What to do when you can't Add and Subtract" (see Appendix).

However, first let me describe a very powerful method for rote learning developed by my friend Dr Colin Lane. As ever, it will not work for everyone, but then, nothing does.

If you are trying to remember a phone number while you look for a pen and paper to write it down you will probably keep repeating it, under your breath. This is sub-vocalising and is the basis of a learning technique called 'self-voice echo'. It seems that we learn best from hearing the information in our own voice, hence 'self voice-echo'. Some years ago now I tested this out on some teenage males who knew very few times table facts, comparing self-voice echo with some other rote learning techniques. The self-voice group made dramatic and long-lasting gains, but even in that group there was one boy who did not make any gain . . . so, the best way to find out if it will work for you is to give it a try.

You can record selected facts onto a PC and also write them on the screen. Put on headphones, which will cut out distracting background noises, and repeatedly 'echo' the fact back through the headphones as you look at the fact on the screen. If you find yourself muttering the fact, then that is even better. It is important to look at the fact as you say it, then you are pushing the fact into your brain by seeing, listening and, if you are mouthing, muttering the fact as you hear yourself (this is the descriptively named sub-vocalising), saying it. This multisensory input helps memorising.

Practise about five facts at each session. Make a session fairly short as this is a method which is quite demanding on your perseverance levels. Try the technique to find out if it is a method which works for you.

Inter-relating the basic facts

In my experience, most people know at least some maths, often more than they realise. Let's apply this to the basic addition facts. There are 121 basic addition facts. When presented together in the addition square, they seem to be a daunting task. However, if you shade in the facts you know (perhaps the 'add 10' facts or the 'add zero' facts) and the ones you can quickly access (perhaps adding on 1 or 2 by quick counting), then you should find that there is only a small number of facts that cause difficulty. So the learning task is not 121, but more likely to be less than 25 facts.

	0	1	2	3	4	5	6	7	8	9	10
0	0	1	2	3	4	5	6	7	8	9	10
1	1	2	3	4	5	6	7	8	9	10	11
2	2	3	4	5	6	7	8	9	10	11	12
3	3	4	5	6	7	8	9	10	11	12	13
4	4	5	6	7	8	9	10	11	12	13	14
5	5	6	7	8	9	10	11	12	13	14	15
6	6	7	8	9	10	11	12	13	14	15	16
7	7	8	9	10	11	12	13	14	15	16	17
8	8	9	10	11	12	13	14	15	16	17	18
9	9	10	11	12	13	14	15	16	17	18	19
10	10	11	12	13	14	15	16	17	18	19	20

One of the essential strategies in this book is to build on and extend the facts and knowledge you already have and thus develop knowledge and understanding. Another essential strategy is to identify and learn the KEY facts, so that you learn the ones that are going to be the most useful.

The addition facts are a good example of this. They also illustrate the idea of finding the 'easy' (or key) number and 'easy' fact within the 'hard' numbers and 'hard' facts. Building on what you know is a good way to learn, particularly if you build in a mathematically developmental way. Let's look at two examples of this idea.

The doubles

One of the addition facts which is commonly known is 5 + 5 = 10

First, add on one more to make 5 + 5 into 5 + 5 + 1 which makes an answer of 11. 5 + 5 has become 5 + 6 (or 6 + 5), so

$$5 + 6 = 11 \qquad \text{and} \qquad 6 + 5 = 11$$

This is 'the doubles plus 1' strategy

Now, a second extension of 5 + 5:

Return to 5 + 5 and move 1 from the first 5 to the second 5. This keeps the answer as 10, but now the two numbers adding to make 10 are;

$$4 + 6 = 10 \qquad \text{and} \qquad 6 + 4 = 10$$

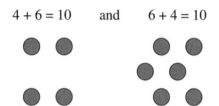

This is 'the sharing strategy'

A third extension of 5 + 5 = 10 takes us to 5 + 4 and 4 + 5 = 9

This is 'the doubles minus 1 strategy'

These three modifications have extended one fact, 5 + 5 = 10 into six more addition facts. The other 'doubles' can be treated the same way.

Adding 9

This example uses the pattern of adding on 10. This pattern is reliable and predictable when it is written in digits:

$$10 + 1 = 11$$
$$10 + 2 = 12$$
$$10 + 3 = 13$$
$$10 + 4 = 14$$
$$10 + 5 = 15$$
$$10 + 6 = 16 \text{ etc}$$

(This is another example of place value at work and, indeed, the sheep counters).

The pattern is less predictable and reliable when expressed in words. For example, eleven and twelve are exceptions to any word pattern used for the 'teen' numbers. The 'teen' pattern is confusing in that the word gives the unit digit first, as in 'four' and the ten digit second, as in 'teen' to give 'fourteen' which we write as a ten followed by a four, 14.

Now, back to linking facts to make more facts. This time by relating 9 to 10. . . .

The key relationship is that 9 is close to 10, 9 is 1 less than 10, so comparing the addition fact of 9 plus 6 to the addition of 10 plus 6, the answer must be smaller. It will be 1 less. This gives the second step in this strategy for adding 9. You subtract 1:

Step 1 $10 + 7 = 17$ Step 2 $17 - 1 = 16$

For some people adding 9 is one hard step. But adding 9 can be done instead by using two easy steps. It may also lead to better accuracy than counting on one at a time.

Step 1)   Add 10
Step 2)   Subtract 1

9 is the hard number, 10 and 1 are the easy numbers, used instead of 9.

These two ideas can be used to access several facts. . . .

Extending the collection of seven 'doubles' shown below leads to 28 other facts (and the 'doubles minus 1' facts, too)

2 + 2 = 4	2 + 3 = 5	3 +2 = 5	1 + 3 = 4	3 + 1 = 4
3 + 3 = 6	3 + 4 = 7	4 + 3 = 7	2 + 4 = 6	4 + 2 = 6
4 + 4 = 8	4 + 5 = 9	5 + 4 =9	3 + 5 = 8	5 + 3 = 8
5 + 5 = 10	5 + 6 = 11	6 + 5 = 11	4 + 6 = 10	6 + 4 = 10
6 + 6 = 12	6 + 7 = 13	7 + 6 = 13	5 + 7 = 12	7 + 5 = 12
7 + 7 = 14	7 + 8 = 15	8 + 7 = 15	6 + 8 = 14	8 + 6 = 14
8 + 8 = 16	8 + 9 = 17	9 + 8 = 17	7 + 9 = 16	9 + 7 = 16

Adding 9 by using adding 10 and subtracting 1 is a 'two easy steps for one hard step' strategy. Adding 8 can be done by a similar idea. Since 8 is 10 take away 2, adding 8 can be done by adding 10 and subtracting or counting back 2.

So, fourteen facts are derived from seven facts;

3 + 10 = 13	3 + 9 = 12	3 + 8 = 11
4 + 10 = 14	4 + 9 = 13	4 + 8 = 12
5 + 10 = 15	5 + 9 = 14	5 + 8 = 13
6 + 10 = 16	6 + 9 = 15	6 + 8 = 14
7 + 10 = 17	7 + 9 = 16	7 + 8 = 15
8 + 10 = 18	8 + 9 = 17	8 + 8 = 16
9 + 10 = 19	9 + 9 = 18	9 + 8 = 17

And 14 more facts are available via the 7 'reverse' facts 10 + 3, 10 + 4 and so on.

Adding or subtracting 10 instead of adding or subtracting 9 is an example of estimation. Whenever you adjust the estimation to obtain the exact answer, there is a useful question, 'Is the estimate bigger or smaller than the accurate answer?'

If you add 10 instead of 9, the estimate is bigger. So this tells that you have to take away 1 to obtain the exact answer.

If you subtract 10 instead of 9, you have taken away more which gives you an estimate that is smaller. So you have to add back 1 to obtain the exact answer.

The number bonds for 10. Two numbers which add to make 10.

Another collection of addition facts which is really useful is the collection which mathematicians call the 'number bonds for 10'. These are the pairs of numbers which add to make 10. They start with 10 + 0 and then follow a pattern through 5 + 5 to 0 + 10.

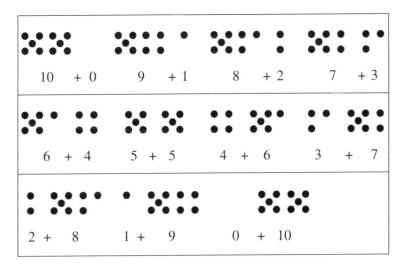

Another way to picture the number bonds for 10 is to use the ten fingers on your hands. As you move your fingers over, one at a time, you move through the number bonds for ten (literally, 'facts at your finger tips').

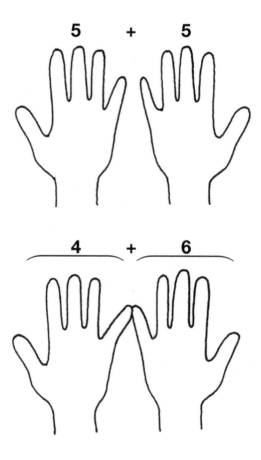

5 + 5 is easy to remember as we have a reminder in front of us all the time . . . the five fingers on each of our two hands.

Subtraction facts

Each addition fact sets up two subtraction facts, for example:

$$6 + 8 = 14 \qquad 14 - 6 = 8 \qquad 14 - 8 = 6$$

Subtraction facts can also be seen as 'What do I add onto x to make y?' That x and y is some algebra. Here it is being used to generalise a process. Two specific examples using numbers are:

What do I add onto 6 to make 14?   and

What do I add onto 8 to make 14?

Times table (basic multiplication) facts

Recently I spoke with a very intelligent young woman who is studying maths at A level. She has been offered a place at Oxford University to study engineering. We were talking about her difficulties with some areas of maths and she recalled with great feeling and self awareness her experiences as a seven year old of being unable to commit the times tables to memory. All the children in the class who succeeded at this task received a special badge. She didn't and although she can look back and see that this seems trivial now, at the time it had an enormous effect on her self-image and her faith in her ability to do maths. Sadly this story is not unique.

One of the problems with this collection of facts is that people think they can teach any child or adult to learn them 'by heart' providing that child or adult puts in enough effort. This is not true and so then maybe people use music, believing that if you put the times tables to a catchy tune, THEN people will learn them. Inevitably this will not be true for everyone. So people use funny stories such as, two sixes are lost in a desert and become 'thirsty sixes' (36?). You have to remember a lot of stories to cover all the basic multiplication facts.

Being unable to commit these facts to memory is probably one of the first experiences of failure in maths for young pupils. This situation may be a major contributor to maths de-motivation. What

I want to suggest is a way that not only gets around this problem, but actually adds extra understanding of maths.

There are 121 times table facts, which can be presented as the traditional columns or tables, for example the four times table;

$$1 \times 4 = 4$$
$$2 \times 4 = 8$$
$$3 \times 4 = 12$$
$$4 \times 4 = 16$$
$$5 \times 4 = 20$$
$$6 \times 4 = 24$$
$$7 \times 4 = 28$$
$$8 \times 4 = 32$$
$$9 \times 4 = 36$$
$$10 \times 4 = 40$$

What is multiplication? Why do we need it?

As you read down this table, you probably say, 'One times four is four. Two times four is eight.' And so forth. The word 'times' is quite abstract. So, you have to know what it means. You could equally say, 'One lot of four is four. Two lots of four are eight,' These words give more meaning. Better still, for understanding, would be, 'Two lots of four added together make eight.' But that is a lot more words than 'Two fours are eight.'

That is what multiplication means. It means adding 'lots' of the same number together. So 5×4 is actually maths shorthand for $4 + 4 + 4 + 4 + 4$

These facts are another example of progression on from counting in ones. It would take a while to count on in ones for 5×4, and the counting may not end up as being accurate.

The times table facts can also be presented, altogether at one time, in a table square. This presentation is useful, but rather daunting in the amount of information it displays at one time.

	0	1	2	3	4	5	6	7	8	9	10
0	0	0	0	0	0	0	0	0	0	0	0
1	0	1	2	3	4	5	6	7	8	9	10
2	0	2	4	6	8	10	12	14	16	18	20
3	0	3	6	9	12	15	18	21	24	27	30
4	0	4	8	12	16	20	24	28	32	36	40
5	0	5	10	15	20	25	30	35	40	45	50
6	0	6	12	18	24	30	36	42	48	54	60
7	0	7	14	21	28	35	42	49	56	63	70
8	0	8	16	24	32	40	48	56	64	72	80
9	0	9	18	27	36	45	54	63	72	81	90
10	0	10	20	30	40	50	60	70	80	90	100

As with the addition facts, it is my experience that very few people know absolutely none of the multiplication facts. People tend to focus on the negative here, on what they don't know. Dealing with these facts requires a positive approach.

Whatever your own feelings of your achievements or otherwise in mathematics you will have learned at least some of these facts, usually the tables with the best patterns, that is the 0×, 1×, 2×, 5× and 10×. Luckily, these are the most important facts to know. Surprisingly, if you use the table square as a way of looking at all the facts at once, filling in these key facts shows that there are only 36 facts to learn. Later on I will show you how to reduce these 36 with very little effort.

The reason why these 121 facts reduce to 36 is that each of those tables with the best patterns contain some of the facts from the harder tables. For example, from the seven times table facts the 'easy' facts are; 0×7, 1×7, 2×7, 5×7 and 10×7.

The number of remaining facts on the table square is quickly reduced even further by a useful maths idea, that is, you get the same answer to a multiplication of two numbers irrespective of which number multiplies which.

For example, 6 × 8 and 8 × 6 both give 48 and 3 × 4 and 4 × 3 both give 12. Two facts from one fact.

To explain this, try setting out 3 × 4 in 1p coins

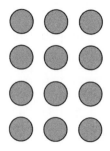

Describing what you have depends on whether you look at rows or columns. There are 4 rows of 3 or 3 columns of 4. Both total 12.

The toughest fact in the times table is generally considered to be 7 × 8 (and the same fact in its 8 × 7 form). This particular fact is one of only two to follow a special pattern

5 6 7 8

becomes

56 = 7 × 8

The other fact is 3 × 4 = 12

1 2 3 4

becomes

12 = 3 × 4

The easy steps strategy

The remaining missing facts can be treated by the same idea we used for addition facts, that is to use what you know to make one hard step into two easy steps. That is, relate the numbers, particularly by finding the 'easy' numbers in the 'hard' numbers.

I shall also remind you of the link between multiplication and addition, since this is key idea for the two easy steps methods.

Multiplication is the repeated addition of the same number. A times table fact such as 6×8 is the repeated addition of six eights.

$8 + 8 + 8 + 8 + 8 + 8 = 48$.

This addition can be done one step at a time, but it is also easy to group or chunk the numbers:

8+ 8 + 8 + 8 + 8 + 8

as shown, the 6×8 has been chunked as 5×8 and 1×8.

So 6×8 is 5×8 (= 40) plus 1×8 (= 8), that is 48.

Another example;

7×4 is $4 + 4 + 4 + 4 + 4 + 4 + 4$

The 4s can be added one at a time or grouped into five 4s and two 4s. This adapts 7×4 making it into two easier steps

$7 \times 4 =$ **4 + 4 + 4 + 4 + 4**   + $4 + 4$

$5 \times 4 = 20$ and $2 \times 4 = 8$,

So $7 \times 4 = 20 + 8 = 28$

This strategy is, again, moving on from counting in ones. By grouping or chunking numbers, the maths process is made more efficient.

So, this strategy involves finding the combinations of easy numbers within the harder number. These are:

$3 = 2 + 1$
$4 = 2 \times 2$
$6 = 5 + 1$
$7 = 5 + 2$
$8 = 2 \times 2 \times 2$ (or $10 - 2$)
$9 = 10 - 1$

Thus, the two steps for calculating times tables facts are;

A 3× fact is calculated as a 2× fact plus a 1× fact (2 + 1 = 3).
A 6× fact is calculated as a 5× fact plus a 1× fact (5 + 1 = 6).
A 7× fact is calculated as a 5× fact plus a 2× fact (5 + 2 = 7).

A 4× fact is calculated as a 2× fact doubled (2 × 2 = 4).
A 8× fact is calculated as a 2× fact doubled and then doubled again (2 × 2 × 2 = 8)

The two step strategy also works with a subtraction.

For example, to take the 10x facts back to the 9x facts:

A 9× fact is calculated as a 10× fact minus a 1× fact (because 9 = 10 − 1, a link we used in Chapter 4).

An 8× fact can (also) be calculated as a 10× fact minus a 2× fact (8 = 10 − 2)

The 9x facts and patterns

The 9× table is a good table for patterns. And patterns help memory. The patterns take 9, which has an image of being difficult, and make it a much easier, friendlier number.

Let's look at the 9× table;

$1 \times 9 = 9$
$2 \times 9 = 18$
$3 \times 9 = 27$ The answers have patterns. If you add
$4 \times 9 = 36$ up the digits in any of the answers,
$5 \times 9 = 45$ the total is always 9, for example:
$6 \times 9 = 54$
$7 \times 9 = 63$ . . . . . . $6 + 3 = 9$
$8 \times 9 = 72$
$9 \times 9 = 81$ . . . . . . $8 + 1 = 9$
$10 \times 9 = 90$

There is another pattern in the answers to the nine times table. If you look down the unit digits in the answer column, starting at the top, you can see how they sequence as 9, 8, 7, 6, 5, 4, 3, 2, 1, 0. The digits in the tens column also sequence, but in the opposite direction, starting with an unwritten zero (in $1 \times 9 = 09$) and ending at 9.

The adding digits pattern extends to all multiplication by 9. The answer to *any number* times 9 will have digits which ultimately add up to make 9. For example,

85 461 $\times$ 9 = 769 149. Add all the digits in the answer, $7 + 6 + 9 + 1 + 4 + 9 = 36$. Then add again, $3 + 6 = 9$.

Now, let's try another pattern in the 9× table. This pattern works beyond just the table facts to any number times 9. Once again, we take a harder number and relate it to an easy number. We will use the 10× facts to take us to 9× facts.

All this work is based on 9 being 1 less than 10, a relationship used on pages 35 for extending basic addition facts and again on page 63 for addition of shop prices. Using 10 to access 9 is using estimation skills.

You may find it helpful to use coins to illustrate the development of this idea. 9p is 1p less than 10p. So it follows that 2 lots of 9p would be 2p less than these 2 ten pence coins. If you took 6 ten pence coins then 6 lots of 9p would be 6p less than the 6 ten pence coins.

Build up the 9× table in comparison with the 10× table.

1 nine is 1 less than 1 ten	$1 \times 9 = 1 \times 10 - 1 = 10 - 1 = 9$
2 nines are 2 less than 2 tens	$2 \times 9 = 2 \times 10 - 2 = 20 - 2 = 18$
3 nines are 3 less than 3 tens	$3 \times 9 = 3 \times 10 - 3 = 30 - 3 = 27$
4 nines are 4 less than 4 tens	$4 \times 9 = 4 \times 10 - 4 = 40 - 4 = 36$
5 nines are 5 less than 5 tens	$5 \times 9 = 5 \times 10 - 5 = 50 - 5 = 45$
6 nines are 6 less than 6 tens	$6 \times 9 = 6 \times 10 - 6 = 60 - 6 = 54$
7 nines are 7 less than 7 tens	$7 \times 9 = 7 \times 10 - 7 = 70 - 7 = 63$
8 nines are 8 less than 8 tens	$8 \times 9 = 8 \times 10 - 8 = 80 - 8 = 72$
9 nines are 9 less than 9 tens	$9 \times 9 = 9 \times 10 - 9 = 90 - 9 = 81$

The pattern is shown in this table. If you need to work out a 9x fact, start with a 10x fact. For example, **6** $\times$ 9 starts with **6** $\times$ 10 equals 60 and then subtracts **6** to make 54.

Again, if you want **8** $\times$ 9, start with **8** $\times$ 10 as 80 and subtract **8** to make 72. (You can check the answer by adding the digits. If they add to 9 you are correct. Here $7 + 2$ does make 9).

This technique works for any number times 9. Try **35** × 9. . . **35** × 10 is 350. Now subtract **35** from 350 to obtain 315. Check by adding the digits, 3 + 1 + 5 = 9.

This method for working out the nine times table facts is a classic example of using two easy steps to achieve one difficult step. Of course if multiplying by nine is not a difficult step for you, you won't need to do this!

This method is also about estimations. Multiplying by 10 instead of 9 gives an estimate of the answer. This estimate will be bigger than the accurate answer. So adjustment to the accurate answer will involve subtraction.

Multiplying and dividing by 4 using two steps

This two step strategy can be used for multiplying (and dividing) by 4. To multiply by 4, you multiply by two, twice, for example, to answer 4 × 7;

multiply 7 by 2 to give 14, then multiply by 2 again to get 28, that is

$7 \times 2 = 14$ followed by $14 \times 2 = 28$

To divide 884 by 4, again use two steps;

Step 1 $884 \div 2 = 442$
Step 2 $442 \div 2 = 221$

Multiplying and dividing by 8 using 3 steps

Since 8 is 2 × 2 × 2, multiplying by 8 can be achieved by multiplying by 2 three times, for example: 7 × 8

First step $7 \times 2 = 14$
Second step $14 \times 2 = 28$
Third step $28 \times 2 = 56$

Dividing by 8 can be achieved by dividing by 2 three times, for example:

$96 \div 8$

First step	$96 \div 2 = 48$
Second step	$48 \div 2 = 24$
Third step	$24 \div 2 = 12$

The square numbers

All four sides of the square have the same length and all four angles are 90° (see page 116).

The area of a square is calculated by the same method as that used to calculate the area of a rectangle, that is to multiply length times width, except that these two lengths are the same in a square. So for a square with sides of length 5, the area is $5 \times 5 = 25$.

If a number is multiplied by itself as in 5×5, the answer is called a square number. The first ten square numbers are;

$$1 \times 1 = 1$$
$$2 \times 2 = 4$$
$$3 \times 3 = 9$$
$$4 \times 4 = 16$$
$$5 \times 5 = 25$$
$$6 \times 6 = 36$$
$$7 \times 7 = 49$$
$$8 \times 8 = 64$$
$$9 \times 9 = 81$$
$$10 \times 10 = 100$$

There is a pattern for this sequence of numbers. The difference between successive numbers gives the pattern:

$$4 - 1 = 3$$
$$9 - 4 = 5$$
$$16 - 9 = 7$$
$$25 - 16 = 9$$
$$36 - 25 = 11$$
$$49 - 36 = 13$$
$$64 - 49 = 15$$
$$81 - 64 = 17$$
$$100 - 81 = 19 \ldots \textit{the odd numbers} \ldots.$$

Mathematicians often use ways to write information as briefly as possible whilst retaining total clarity. Of course, the total clarity is only available if you know the code. The code for squares is a 2, but written in a special place and in a smaller font, for example:

$$5 \times 5 = 5^2$$
$$8 \times 8 = 8^2$$
$$10 \times 10 = 10^2$$

This can be written as a generalisation by using algebra and using a y to represent any number:

$$y \times y = y^2$$

5 Add, Subtract, Multiply and Divide

Mathematicians call these the four operations. They are the four basic ways you manipulate (operate on) numbers and so it is very important that you spend time making sure you have a really good and clear understanding of what each operation means and also how they all relate to and link with each other. This work is very much the maths that is 'everyday' (but it also sets the foundations for many future topics).

Adding

Most people picture addition as 'sums' such like these;

$$5 + 6 \qquad 29 + 37 \qquad \begin{array}{r} 436 \\ +278 \\ \hline \end{array}$$

but adding starts with counting.

Counting 1, 2, 3, 4, 5, 6, 7 and so on is adding 1 each time.

Counting 2, 4, 6, 8 is adding 2 each time.

Counting 10, 20, 30, 40, 50 is adding 10 each time.

Adding is putting together. For example, putting together 5 coins and 3 coins gives a total of 8 coins. Addition sums such as;

$$\begin{array}{r} 56 \\ +23 \\ \hline 79 \end{array}$$

involve putting together two numbers (56 and 23) to make a total of 79.

Subtracting

Subtracting usually conjures up images of sums such as;

$$9 - 3 \qquad 46 - 31 \qquad \begin{array}{r} 787 \\ -463 \\ \hline \end{array}$$

Subtraction is taking away, separating and therefore it is the opposite of adding. This means that counting back 10, 9, 8, 7, 6 is subtracting one each time.

Counting back 8, 6, 4, 2 is subtracting two each time.

Counting back 70, 60, 50, 40, 30 is subtracting ten each time.

The taking away and separating aspects of subtraction are illustrated in the following examples:

If you had 8 coins and you take 3 away you have 5 left. You have separated the 8 coins into 3 coins and 5 coins.
Subtraction problems like;

$$\begin{array}{r} 98 \\ -46 \\ \hline 52 \end{array}$$

involve separating the number 98 into two parts. The number 46 is one part. By taking away 46 from 98 you find the second part, 52.

Multiplication

Multiplication is repeated addition, or more precisely, adding the same number several times. For example 7×6 is seven sixes added together, $6 + 6 + 6 + 6 + 6 + 6 + 6$. Although it would be possible, and mathematically correct to work out 7×6 in this adding way, it is quicker to know the answer is 42 or to work it out by a more efficient method than six additions. For example, you can combine adding and multiplying so that 7×6 is broken down into 5×6 plus 2×6. (See page 44)

It is very important to realise that multiplication is so closely related to adding. That it is really a special form of addition.

The method most people learned at school, called 'long multi-plication', is a mixture of addition and multiplication. I will explain this in more detail on page 67.

Division

Division is dividing up into equal parts. It is the opposite of multi-plication. This means it is repeated subtraction, taking away the same number several times. For example, to divide 36 by 9, you could take away 9's successively until you reached zero;

$$36 - 9 = 27$$
$$27 - 9 = 18$$
$$18 - 9 = \ 9$$
$$\ 9 - 9 = \ 0$$

Four 9's were subtracted. $36 \div 9 = 4$ This could also be expressed as 'How many nines are there in 36?

I hope you can see how the four operations, add, subtract, multiply and divide are so closely related and how they can be mixed and blended to suit the way you want to work.

One of the big misconceptions about mathematics is that there is only one way to do something. It is possible to use ideas such as the link between addition and multiplication to provide alternate methods for calculating.

The traditional methods for addition and subtraction

These are the written methods which involve 'carrying', 'borrow-ing', 'renaming', 'regrouping' and 'decomposing'. These words are a good illustration of the point I made on page 7 about the lan-guage of maths. They all have other meanings which have nothing to do with maths. This handicaps the clarity of their use to com-municate maths concepts, in that it is less likely learners will automatically attach an image, a picture or visual meaning to these words that will link them to maths.

Never-the-less, I shall now embark on explanations which will use these words (and some alternatives, which in my biased opinion are better). I will work through, in detail, the addition of two numbers to get a total and then do the same for a subtraction using the same numbers, taking the total apart again. This should illustrate once again that addition and subtraction are the same processes in reverse. It might even make the 'carrying' and 'decomposing' clearer.

So, even if you think you are an expert at addition, humour me and work through the explanation.

As each example unfolds, look for the strong links to counting on and back, especially crossing the tens and hundreds (remember the sheep counters and their nudges?)

Addition

The addition example, which may be written in two ways, is:

$$187 + 269 \qquad \begin{array}{r} 187 \\ +269 \\ \hline \end{array}$$

To help you develop a clearer picture of each step in the procedure, try using 1p coins (for units), 10p coins (for tens) and £1 coins (for hundreds) to illustrate what is happening.

Set up the coins.

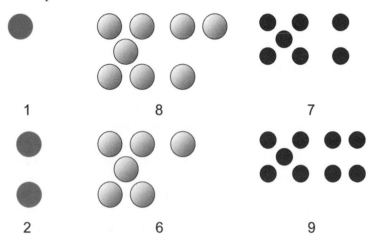

In the written method you start at the units. Since addition is putting together, move the 7 one pence coins and 9 one pence coins together. This gives 16 coins.

Now exchange (trade) ten 1p coins for one 10p coin.

Trading or exchanging ten units for one ten, one ten for ten units, ten tens for one hundred, one hundred for ten tens and so on is a fundamental procedure in arithmetic. I always think of it in terms of real life when I try to reduce the number of coins I carry in my pocket by exchanging, for example, ten 10p coins for a £1 coin.

16 is a two digit number, 1 (ten) and 6 (units). The 6 stays in the units column and the ten is (logically) moved to the tens column. This is the "carry" number.

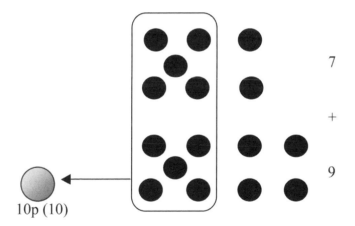

10p (10)

7

+

9

The next step is to move to the tens column:

and put the 8 ten pence, 6 ten pence and the 1 carried ten pence coins together. This gives 15 ten pence coins.

Now exchange (trade) ten 10p coins for £1 coin.

Now you have a £1 coin and five 10p coins. The 5 ten pence coins stay in the tens column and the £1 coin (which represents 100) is moved (carried) to the hundreds column.

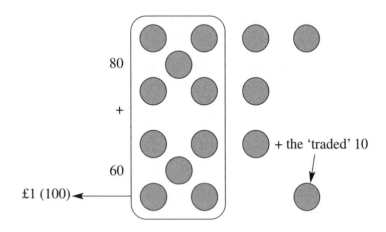

80

+

60

£1 (100)

+ the 'traded' 10

All that remains is to put together the hundreds, the £1 coins. Add together the 1 (100) and the 5 (500) from the sum and add in the carried 1 (100) to make 7 (700). The addition is complete.

$$\begin{array}{r} 187 \\ +569 \\ \hline 756 \end{array}$$

The coins follow exactly the same process as the written method. Coins can be useful to illustrate the process and make the symbols (digits) have some meaning.

Subtraction
The subtraction may be written in two ways:

$$756 - 187 \qquad \text{or} \qquad \begin{array}{r} 756 \\ -187 \\ \hline \end{array}$$

The subtraction example uses the same numbers we used in the addition example and separates the total into its two parts again.

756	is the total
−187	is one part
	and the answer will be the other part

What the question actually "says" is:

"Here is 756, take away 187 and find out what is left."

Using coins to work through the question will make each step clear, because the coin movements mirror the written method.

Set up 756 in writing and in coins (seven £1 coins, five ten pence coins and 6 one pence coins).

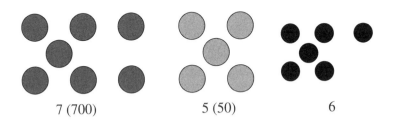

7 (700)   5 (50)   6

The subtraction starts with the units (the one pence coins). You have to take away 7 coins from the 6. Obviously 6 is smaller than 7, so this makes this first step in the subtraction a little tricky without some adjustment.

To obtain more 1p coins, you reverse the 'carrying' you did for addition. You have to exchange (or trade) one 10p coin for ten 1p coins. This takes the number of ten pence coins in the tens place value column down to 4 (from 5) and puts 16 one pence coins in the units place. School textbooks sometimes call this 'decomposition' because you have decomposed the 756 into six hundreds, four tens and sixteen units. It may also be called 'renaming' (because you have done that, too).

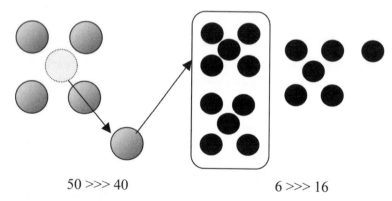

50 >>> 40   6 >>> 16

Now you can take away 7 one pence coins, leaving 9 one pence coins. (The 9 is the units digit of your answer). Move on to the tens . . .

You have 4 tens and you need to take away 8.

As 4 is less than 8, again you have to trade (decompose the number).

Again do the opposite trade you did in the addition. Trade one of the £1 (hundred) coins for ten 10p coins. This leaves 6 hundreds (six £1 coins).

Move the 10p coins in the tens column to make 14 ten pence coins altogether.

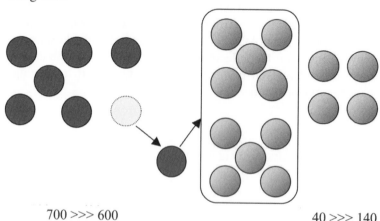

700 >>> 600   40 >>> 140

Now it is possible to take away 8 tens, leaving 6 as the tens digit of your answer.

The last step is to subtract in the hundreds column. There are 6 hundreds (six £1 coins) left.

You take away 1 to leave 5 hundreds as your answer in the hundreds place value column.

Looking back over the procedure, the complete answer is 569, which, of course is the number you added to 187 to make 456.

This decomposition method breaks down a number into parts which are more suitable for the subtraction. Breaking down numbers is a technique that is often useful in maths. This particular breaking down technique is aimed to produce enough units, or tens, or hundreds, etc to make a subtraction possible.

In 756 the units and the tens figures were not large enough for the subtraction. 756 was decomposed or renamed as:

6 hundreds, 14 tens and 16 units

$$
\begin{array}{lll}
\text{This is still } 756 \ldots\ldots\ldots & 600 & \text{6 hundreds} \\
& 140 & \text{14 tens} \\
& +\,16 & \text{16 units} \\
\hline
& 756 &
\end{array}
$$

This is another example of maths using the strategy of breaking down and building up numbers in order to make a calculation easier.

Here are two ways to write down this method of subtraction;

$$
\text{the messy} \quad
\begin{array}{c}
{}^{6\ \ 14\ 16} \\
7\ \ 5\ \ 6 \\
-1\ \ 8\ \ 7 \\
\hline
\end{array}
\qquad \text{or the middle line} \quad
\begin{array}{c}
7\ \ 5\ \ 6 \\
6\ 14\ 16 \\
-1\ \ 8\ \ 7 \\
\hline
5\ \ 6\ \ 9
\end{array}
$$

(if you think that this is a complicated explanation, try listening to Tom Lehrer singing about New Math . . . just Google 'Tom Lehrer New Math').

Doing subtraction 'in your head'

The written method described above would ask a great deal of your working memory, the memory you will use while working out this subtraction, holding the figures in your mind while you move through each step. There is another disadvantage in the method above. In this written method you work from units to hundreds, so you move across to the left, having to hold the unit and tens digits in your brain, with no practise or repetition as you move on to new digits, which may well confuse your memory's hold on the earlier digits. The final answer presents these digits in the reverse order to the way in which you did the calculation.

If you work in the direction from hundreds to units, you can repeat the digits in the evolving answer as you work along.

The 'hundreds to units' method.

The explanation takes a lot longer than the method!

> 756 – 187.

Start at the hundreds;
> 7 (hundred) minus 1 (hundred) = 6 (hundred)

Move to tens:
> 5 (tens) minus 8 (tens) . . . this needs some help, so:

A nudge (back to those sheep counters), which takes one hundred from the 6 hundreds and changes it into 10 tens. This leaves 5 hundred, but makes the 5 tens up to 15 tens.

Then
> 15 (tens) minus 8 (tens) = 7 (tens)

The answer, so far is five hundred and seventy . . . .

Move to units:

> 6 (units) minus 7 (units) . . . this needs some help, so

A nudge takes 1 ten from the seven tens, leaving the tens digit as sixty and providing an extra ten units, making the 6 units up to 16.

Then

16 (units) minus 7 (units) = 9 (units)

The final answer is 569

The 'adding on' method.

Now try the' adding on' method, which once again uses the fact that addition and subtraction are reversals of the same idea.

756 – 187

The principle is to start at 187 and add on numbers until you reach the target number of 756. Use sensible stages (go for tens, hundreds, thousands and so on).

187 plus 3 is 190	3
190 plus 10 is 200 . . . so far 13 added on	13
200 plus 500 is 400 . . . so far 513 added on	513
513 plus 56 is 169 . . . 513 + 56 is 569 added on	569

You could add the 56 in two stages, first 50, then 6.

This method uses a running total. The repetition of the numbers at each stage helps the working memory retain the numbers.

The 'estimate and refine' method.

Finally there is the method that starts with an estimate and then refines it. In this example, 756 -187, the 187 can be rounded to 200 as a first step.

Then subtract   756 – 200 = 556   this is the estimate answer.

Since you added 13 to 187 to make 200, your estimate is smaller than the accurate answer. You have to add back the 13.

556 + 13 = 569

Having a choice of methods allows you to select the method that works best for you. Also, seeing each each method may add to your understanding of the processes.

Two ways to add a column of numbers.

This is a difficult task for many people. Yet it is a task that occurs quite often in life, with shopping bills and restaurant and cafe bills, for example.

If people use a calculator, they often make a mistake at one stage and get the total wrong. The two methods described below are quite different from each other, but both provide real support for the adding process.

1. The tally method

(Again, the description is longer than the method. This particular method tends to appeal to inchworms rather than grasshoppers).

```
  46
  78
  65
  93
  28
  57
  44
 +38
 ————
 449
```

Start at the top of the units column and add $6 + 8 = 14$. Draw a slash through the 8 (this tally represents the ten from 14) and carry on adding down with the 4 (from the 14) added to the 5 to make 9. Next digit: $3 + 9 = 12$. Draw a slash through the 3 (this tally represents the ten from the 12) and carry on adding down, $2 + 8 = 10$. Draw a slash through the 8 to represent another ten. Add on down $7 + 4 = 11$. Another slash, through the 4. Add on down $1 + 8 = 9$. The 9 goes in the units column as its total.

Now count the tally marks (slashes). There are four, 4, tallies. This represents 4 tens to write at the top of the tens column.

Now add down the tens column using the same tally method, but each tally mark is now worth one hundred, 100.

4 + 4 = 8. Then 8 + 7 = 15, so a slash through the 7 represents a hundred and the 5 is carried on. 5 + 6 = 11. A slash through the 6 represents another 100.

Carry on adding down the column. 1 + 9 = 10. Another slash for another 100, through the 9. Adding down again, 2 + 5 = 7. Then 7 + 4 = 11. Draw a slash for another 100 through the 4.

1 + 3 = 4. Write 4 as the total of the tens column.

Count the slashes / tallies for the 100's. There are four, 4.

Write 4 as the total in the hundreds column.

This makes the final answer 449.

You can make an estimate to get a rough check on your answer by over-viewing the numbers. There are 8 numbers. Take an average value of 50. The estimate is 8 × 50 = 400.

2. The casting out tens method

```
  42
  95
  67
  24
  33
  56
  98
  33
 +51
 499
```

In this method you look within each column for pairs or triplets of digits which add up to 10 or 20.

From the units column,
2 + 5 + 3 = 10 . . . . cross out 2, 5 and 3
7 + 3 = 10 . . . . cross out 7 and 3   There are 3 tens
4 + 6 = 10 . . . . cross out 4 and 6

All that is left is 8 + 1 = 9.

> 9 is the total for the units column.

The 3 from the 3 tens is written at the top of the tens column.

From the tens column,

> 4 + 6 = 10 . . . . cross out 4 and 6
> 9 + 2 + 9 = 20 . . . . cross out 9, 2 and 9
> 5 + 5 = 10 . . . . cross out 5 and 5

There are 4 tens from the tens column. These 'tens' are actually 100's.

All that is left is 3 + 3 + 3 = 9.

9 is the total for the tens column.

The 4 from the 4 hundreds is written in the hundreds column.

The answer is 499.

Dealing with number 9 and 99 and 999 and 97 and £12.95 and all the other numbers which are a bit less than 10, 100 and 1000.

Inchworms do not like 9, 99 and 999 and similar numbers in this range, such as 95 or 997. If they are or have been inchworms who rely on finger counting, these numbers will represent a lot of fingers.

Grasshoppers quite like these numbers. The grasshopper sees relationships in numbers. Grasshoppers relate them to other, usually easier numbers. So they 'see' that 9 is close to 10, 99 and 95 are close to 100, 999 and 997 are close to 1000. With money they 'see' £999 as close to £1000 and £14.99 as close to £15.00.

The grasshopper will have a better time with mental arithmetic which involves these numbers. Inchworm examination setters will rub their hands in sadistic glee as they pile on the 9's in the questions they see as difficult (for other inchworms).

For mental arithmetic questions involving these numbers, it is worth learning some grasshopper techniques. Inchworms will by their nature try to use their written methods to do mental arithmetic and many of these methods will make too much demand on their (working) memory for them to succeed. It is hard for an inchworm to adopt grasshopper methods (and vice-versa), but it is worth them persevering (and overcoming the feeling that if it is easier they must be cheating).

Inchworms also tend to 'see' numbers literally, exactly as they are written. For example, faced with a petrol price of 159.9p per litre they will read this as 159p and not round it up to 160p, which is a more realistic reading. But it requires the inchworm to actually change 159 into 160.

To illustrate this let's use a money example:

Add these values £9.95 + £4.95 + £22.99 + £1.97

I will explain how an inchworm and a grasshopper solve this addition problem.

The inchworm would rather do this sum on paper. If he is trying to do the calculation in his head then he has to visualise the question as he would write it:

    £ 9.95
    £ 4.95
    £22.99
    £ 1.97
    ———————

Inchworms tend not to over-view. This is not good practice!

The numbers are then added down (or up) the column, first the pence, then the ten pence column, then the pound columns, to achieve, hopefully, an answer of £39.86

The grasshopper rounds each price to the nearest pound, so

    £ 9.95 becomes £10.00 (adding 5p)
    £ 4.95 becomes £ 5.00 (adding 5p)
    £22.99 becomes £23.00 (adding 1p)
    £ 1.97 becomes £ 2.00 (adding 3p)

The preliminary total is found by adding the pounds

    10 + 5 + 23 + 2 = £40

This is bigger than the accurate answer, so the 'added on' pence are added up

$5 + 5 + 1 + 3 = 14$p

And subtracted from the £40.00 to give £39.86

This example shows how the numbers can be simplified to give a reduced number of digits so that, if trying to work this out mentally, the load on memory is greatly reduced. If just the first step in the grasshopper method is used, then it provides an estimate for the money spent.

Another grasshopper skill is the clustering or pairing of less easy numbers into easy numbers, for example ten (see page 61 for another use of this strategy).

The following example appears in a GCSE maths textbook. "Add together mentally, 9, 2, 3 and 6." The book then goes on to add them one at a time in the order given.

If you look at these numbers, that is take time to overview, the 3 and 6 make 9, the 2 can be split into 1 and 1, making the two 9's into two tens and the answer into 20.

```
  9     2     3     6
                  9
    10     10
       20
```

Multiplying by 10, 100, 1000, 10 000 and more

This is a useful and easy process. Having said that, it is often not well understood and nor accurately carried out. I claim that it is easy because it follows a simple pattern. First I will show you the pattern. Then I'll explain the maths behind the pattern.

$43 \times 10 = 430$
$43 \times 100 = 4300$
$43 \times 1000 = 43\ 000$
$43 \times 10\ 000 = 430\ 000$
$43 \times 100\ 000 = 4\ 300\ 000$

The pattern is in the zeros. In this example, the number of zeros on each side of the equals sign are the same.

The pattern is, like any good pattern, predictable. When you multiply by 100, with its two zeros, these two zeros appear in the answer, pushing place values up by 100 times, that is two place values. When you multiply by 1000, with its three zeros, these three zeros appear in the answer, pushing place values up by 1000 times.

The maths is simple (trust me). It is basically about place value.

If a number is multiplied by 10 it gets 10 times bigger. If a number is multiplied by 100 it gets 100 times bigger. If a number is multiplied by 1000 it gets 1000 times bigger. And so on.

How do we know a number is 10 times bigger, or 100 times bigger, or 1000 times bigger, or so on?

The main point is that the digits in the number are the same and in the same sequence, but the place values of the digits in the number have increased. For example, if you select the 4 from the 43. In 43, the 4 represents 40. The 4 has a tens place value.

> In $43 \times 10 = 430$, the 4 moves from a tens place value to a hundreds place value.

> In $43 \times 100 = 4300$, the 4 has moved from a tens place value to a thousands place value.

> In $43 \times 1000 = 43\,000$, the 4 has moved from a tens place value to a ten thousands place value.

After the section on decimal numbers we can look at dividing by 10, 100, 1000, 10 000 and so on.

Long multiplication, a mixture of addition and multiplication

Somewhere, sometime, someone probably explained long multiplication to you. You were probably too young at the time to appreciate the structure of the procedure. I always felt it had a 12 Certificate, but was forced onto a PG audience.

Short multiplication is one step, for example $10 \times 5 = 50$ is definitely one step. $16 \times 2 = 32$ is one step for most, but. . . .

Long multiplication is definitely more than one step, but it is just another version of the strategy I described for working out 7×4 (page 43). For 7×4, the multiplication was related to repeated addition, addition of 4 seven times.

$$
\begin{array}{r}
4 \\
4 \\
4 \\
4 \\
4 \\
4 \\
+4 \\
\hline
28
\end{array}
$$

To avoid a step by step addition of $4 + 4 + 4 + 4 + 4 + 4 + 4$, the sum can be grouped as 5×4 plus 2×4, which is a mixture of addition and multiplication.

This technique can be extended to other multiplications, for example,

12×7, which is $7 + 7 + 7 + 7 + 7 + 7 + 7 + 7 + 7 + 7 + 7 + 7$

The twelve 7's can be clustered into two groups, one of ten 7's and one of two 7's

$7 + 7 + 7 + 7 + 7 + 7 + 7 + 7 + 7 + 7 \quad + \quad 7 + 7$

So 12×7 can be grouped as 10×7 plus 2×7

The multiplication has been broken down into two easier parts, $10 \times 7 \ (=70)$ and $2 \times 7 \ (=14)$. The answers to the two parts are added $(70 + 14)$ to give the answer as 84.

The mathematical term for these parts is 'partial products'. . . quite logical!

This procedure extends into 'long multiplication'. Instead of trying one very difficult step, in this procedure the multiplication is broken down into manageable parts. The traditional way of breaking down the numbers follows a fixed procedure. It is usual to split the numbers according to the place values of the digits in the number. For example, to multiply 48 by 35, the split for the 35 is to

30 and 5, so you multiply 48 by 30 and 48 by 5 and then add the two parts together. To multiply 521 by 426, the split is to do one multiplication of 521 by 400, then a second multiplication of 521 by 20 and a third multiplication of 521 by 6. These three multiplication partial products are then added to show the answer to a multiplication 521 by 426.

An example; 48×35

$$48 \times 30 = 1440 \qquad 48 \times 5 = 240$$
$$1440 + 240 = 1680$$

48×30 and 48×5 are called 'partial products' because a product is the answer to any multiplication of number A $\times$ number B and in 48×35, 48×5 is only a partial contribution to the final product.

Note the $\times 30$ partial product and the place values. 30 is three tens, so the partial product is not 144, but 1440.

This is traditionally set out as shown below;

```
   48
 × 35
 1440    this is 48 × 30
  240    this is 48 ×  5   these two 'partial products' are added
 1680    this is 48 × 35
```

I have found that a visual image can sometimes help with understanding this multiplication. A sum like 48×35 is the same as a calculation to work out the area of a rectangle with sides of 48 and 35.

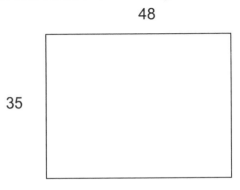

48

35

If the rectangle is broken down into two sub-areas (representing the partial products), using the place value split into units and tens for 35, you have a picture of the two parts of the multiplication.

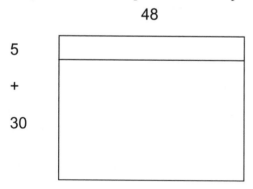

It is, of course, also possible to split a rectangle into four sub-areas.

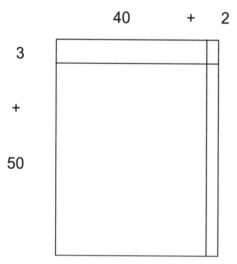

The multiplication of the numbers 42 × 53 can be done in four parts, one part for each of the sub areas of the rectangle.

42×53

a) $40 \times 50 = 2000$
b) $40 \times \ 3 = \ \ 120$ The four parts are added to make the answer.
c) $\ \ 2 \times 50 = \ \ 100$
d) $\underline{\ \ 2 \times \ \ 3 = \ \ \ \ \ 6}$
$\ \ \ \ 42 \times 53 = 2226$

Using a rectangle to represent a two figure number times a two figure number multiplication provides a picture, an image, that can be used again, for example in algebra. So many times in maths, an idea is recycled in different disguises, but it is still the same idea. That's why it helps to understand the first time you meet an idea, even if you don't know where it is leading. It also illustrates how each new interpretation of an idea can make that idea stronger in your mind.

Now, a longer example of long multiplication;

$$521 \times 426$$

In place value terms,   426 is 400 + 20 + 6

This will give us three partial products
(note the place values for × 400 and for × 20)

$521 \times 400 = 208\ 400 \quad 521 \times 20 = 10420 \quad 521 \times 6 = 3126$
$208\ 400 + 10\ 420 + 3\ 126 = 221946$

This is traditionally set out as shown below;

```
    521
  × 426
  208400        this is 521 × 400
   10420        this is 521 ×  20
    3126        this is 521 ×   6
  221946        this is 521 × 426
```

There are occasions when you might combine subtraction and multiplication, for example with a multiplication like 475 × 299. Rather than break down 299 into 200, 90 and 9, it is easier to break 299 into 300 and 1 and subtract (because 300 – 1 = 299).

$$475 \times 300 = 142\ 500 \qquad 475 \times 1 = 475$$
$$475 \times 299 = 142\ 500 - 475 = 142\ 025$$

Indices

Indices are another bit of maths code. Look at these two examples;
the area of a square which has sides of length 5 is 5×5

5×5 is also written as 5^2 (see also page 48)

This is also named as 'five squared' or 'five to the power of 2'

the volume of a cube which has sides of length 5 is $5 \times 5 \times 5$

$5 \times 5 \times 5$ is also written as 5^3

This is also called 'five cubed' or 'five to the power of 3'

It is logical to follow this pattern and write

$5 \times 5 \times 5 \times 5$ as 5^4

This is also expressed as 'five to the power of four'

Indices are useful with powers of ten. This time the logic of the
sequential pattern can be taken forwards and backwards. First the
forwards pattern;

$10 \times 10 = 10^2 = 100$
$10 \times 10 \times 10 = 10^3 = 1000$ (one thousand)
$10 \times 10 \times 10 \times 10 = 10^4 = 10\ 000$
$10 \times 10 \times 10 \times 10 \times 10 = 10^5 = 100\ 000$
$10 \times 10 \times 10 \times 10 \times 10 \times 10 = 10^6 = 1\ 000\ 000$

(one thousand thousand
which is one million)

Now the backwards pattern;

$1000 = 10 \times 10 \times 10 = 10^3$

$100 = 10 \times 10 = 10^2$

$10 = 10 = 10^1$

$1 = 10^0$

$0.1 = \dfrac{1}{10} = 10^{-1}$

$0.01 = \dfrac{1}{100} = \dfrac{1}{10 \times 10} = 10^{-2}$

$0.001 = \dfrac{1}{1000} = \dfrac{1}{10 \times 10 \times 10} = 10^{-3}$

Division the easier way

The procedure called long division, such as $24\overline{)12744}$, is generally considered to be difficult. If long multiplication has a 12 certificate, long division, in its traditional format has a 15, verging on 18 certificate. It certainly requires you to use a lot of maths sub skills. Of course, you can use a calculator, but if you do you should always be able to do at least a rough estimate to check your answer.

Division on a calculator often results in people using the keys in the wrong order.

For example, $24\overline{)12744}$ is keyed in correctly as

Step 1   12744
Step 2   ÷
Step 3   24
Step 4   =

This order for the numbers is a change in the order in which the problem was presented.

There is an alternative method (which also happens to have a built in estimate). I give this a PG, maybe 12 certificate.

The method is based on multiplication and division being the same, but opposite. A basic example should explain this. . . .

Seven times nine equals sixty three $7 \times 9 = 63$

This fact also tells us that there are seven nines in sixty three which is one of the ways of saying that sixty three divided by nine is seven.

$63 \div 9 = 7$

(It also tells us that sixty three divided by seven is nine.

$63 \div 7 = 9$

because every multiplication fact gives us two division facts).

So if 63 is to be divided by 9, you could look at this sum as a multiplication with one number missing . . . the answer.

$63 = 9 \times ?$

This missing multiplier is the basis of the following method for division.

for example $24\overline{)12744}$

    dividing number   original number

In this method you are finding out what number multiplies the dividing number to give the original number, so the example becomes

    $12744 = 24 \times ?$

The division has been turned into a multiplication, another example of the usefulness of rewording maths questions. This is the first stage of this alternate method for division.

The next stage is to set up a simple table. The table sets up easy multiples of 24. Look for the pattern in the numbers in the table. See how the figures 24, 48 and 12 keep recurring?

The number of zeros shows when you have multiplied by 10, 100 or 1000 times 24

1×24	24
2×24	48
5×24	120
10×24	240
20×24	480
50×24	1200
100×24	2400
200×24	4800
500×24	12000

Work on building the table can stop at this point, because 12 000 is close to the number we are dividing. We now have enough information to work out an estimate for our answer.

It should be obvious that the answer is going to be a little bigger than 500, since 500×24 is 12 000, which is close to 12 744, but smaller.

To get the accurate answer, all you have to do is a few subtractions, using the multiplication facts from the table and keeping count of how many 24's you have taken away. Keep the layout organised. . . .

```
  12744
- 12000     500     × 24
    744
-   480      20     × 24
    264
-   240      10     × 24
     24
-    24       1     × 24
      0     531     × 24
```

The answer is 531   531 × 24 = 12 744

Try another example . . . 3960 ÷ 24 . . . ? × 24 = 3960

A look down the 24x table shows the answer is between 100 and 200, closer to 200.

```
  3960
- 2400     100     × 24
  1560
- 1200      50     × 24
   360
-  240      10     × 24
   120
-  120       5     × 24
     0     165
```

The answer to ? × 24 = 3960 (and 3960 ÷ 24) is 165.

6 Estimation

Most of what I have covered so far has been about working out precise answers. But maths, particularly in everyday life, can also be about estimation, where an approximate answer is the goal.

Estimation is a useful approach to numbers, sometimes for checking to make sure that answers are in the 'ball park', that is, not too far wrong to cause chaos. Sometimes an estimate is enough in itself, for example, when working out a tip at a restaurant, or setting a budget for a day out. It's about not needing to be spot-on accurate, but equally about not being miles away from the accurate answer.

In order to estimate you have to be able to relate numbers to key, easier to use numbers, for example, rounding 9 up to 10 or £998 to £1000, or using 5 as half of 10 and 4 as 2 × 2.

Estimates are flexible. They can be close, as in using 100 for 97, or less close, as in using 1000 for 1244.

I like the question, 'Is it bigger or smaller?' for estimation (and other topics in maths, too).

7 Fractions

If I wanted to create maths anxiety I would pick on fractions (and probably division as well). For many people these topics seem to be the start of their disenchantment with maths.

I think the reason is that fractions seem to have their own rules. Whatever logic learners had struggled to achieve up to meeting fractions seems to be destroyed by these rather different numbers.

In the first two editions, this Chapter was much longer. I am now treating fractions as an illustration of how maths can confuse. Also as to how some maths that is taught in school may never be used again. Despite this there are some useful lessons about maths to be learned.

Some fractions are all around us and are part of our everyday life. These tend to be just three, a half, a quarter and a third.

Getting the picture

Where do you meet fractions? Which ones do you meet?
Here are a few common examples.

TIME

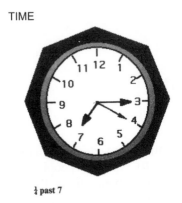

¼ past 7

HAPPY 8 ¾ BIRTHDAY

8 and ¾

Most people know what a half and a quarter mean as we meet these fractions more than any others, especially since the UK "went metric" and eliminated a lot of other fractions based on 'old' units like 16 ounces and 36 inches. If you can understand half and quarter then you can take this understanding, should you wish, to other fractions, because, like all mathematics, rules are rules and all fractions will behave in the same way.

Let's look at half and quarter, written as $\frac{1}{2}$ and $\frac{1}{4}$ respectively.

Firstly $\frac{1}{2}$ uses two digits (1 and 2) with the 1 written on top of the 2 and a line to separate them.

The same is true of $\frac{1}{4}$ or $\frac{3}{4}$ or any other fraction such as $\frac{1}{10}$

These fractions use two digits, but not in the way that, say, twenty-three uses two digits. The two digits in 23 are about place value. This is absolutely not the case with $\frac{1}{2}$ and $\frac{1}{4}$

Now, because a half, a third and a quarter are well known they have developed names which are not in the normal pattern of fraction names. If ½ were to follow the pattern it should be one twoth, ⅓ should be one threeth and ¼ should be one fourth (as it sometimes is in the USA). Subsequently, ⅕ is a fifth, ⅙ is a sixth, ⅐ is a seventh and so on.

So, direct your attention to the digits in the three 'everyday' fractions, and the line that separates them, and they will lead you to an understanding of all fractions.

How a fraction is "made"

I want you to start by using something you can handle to illustrate the maths, a piece of paper, say a sheet of A4.

Take this sheet of paper. This will be your one whole (one sheet). . . . 1.

Now fold it exactly in two. This creates two halves, (which must be exactly the same size). There are now two, 2, parts altogether. One, 1, of these parts is one half, one out of two parts, ½. The 1 sheet has been *divided* into 2 parts

In the fraction $\qquad \frac{1}{2}$

the line under the 1 means divided. The 2 is the number of parts the whole has been divided into and the 1 is the number of parts you have.

There is a link between the divide sign and a fraction

The digits (3 and 4) hide part of the ÷ symbol.

So, the whole has been divided into 2 parts and you have 1 of these parts. . . .1 out of 2 . . . a half ½.

Now fold the paper in half again, which creates four quarters. Each one of these parts is one quarter, one out of four parts, ¼.

In the fraction $\frac{1}{4}$

this line under the 1 means divided. The 4 is the number of parts the whole has been divided into and the 1 is the number of parts you have.

So, the whole has been *divided* into 4 parts and you have 1 of them . . . a quarter, ¼.

This will work with any fraction, for example ¾

$$\frac{3}{4}$$

The 4 and the line mean the whole has been divided into 4 parts. The 3 means that you have 3 of these 4 parts . . . three fourths or three quarters.

Make ¾ with a sheet of paper. Fold it in half one way and then fold this folded paper in half again the other way. This gives 4 equal parts. Cut off one quarter to leave ¾.

We can use ¾ with time, for example, ¾ of an hour. . . .

One hour is 60 minutes.

One quarter, ¼ of an hour is 60 minutes divided by 4, that is 15 minutes.

Three quarters, of an hour is 3 × one quarter, that is 3 × 15, that is 45 minutes.

We can use ¾ with age, for example, ¾ of a year. . . .

One year is 12 months. To find ¼ of a year we divide 12 by 4 to give 3 months.

If 3 months is one quarter then three quarters of a year must be 3 × 3, that is 9 months.

Equivalent fractions: There is more than one way to write a fraction.

There is more than one way of writing a half in digits. It doesn't always have to be ½.

Some other versions of a half are:

Think about half of a year, half of 12 months? This is ⁶⁄₁₂ . . . you have 6 out of 12 months.

What about half of £1? This is 50 pence out of 100 pence. This is ⁵⁰⁄₁₀₀ . . . you have 50 out of 100 pence.

What about half an hour, half of 60 minutes? This is ³⁰⁄₆₀ . . . you have 30 out of 60 minutes.

So far we have four ways of writing a half as fraction numbers

$$\frac{1}{2} \qquad \frac{6}{12} \qquad \frac{50}{100} \qquad \frac{30}{60}$$

What connects these four ways together, what makes them all a half is:

In each of these versions of a half, the top number is half the value of the bottom number. Each is 'equivalent', another version of the half, but still a half.

What a half, a third and a quarter tell us about fractions is:

Always consider the bottom number first. It tells you how many parts in the fraction and therefore the name of the fraction. So, for example, in $\frac{5}{7}$ the bottom number is 7, there are a total of seven parts and the fraction's name is sevenths.

This bottom number tells us how many parts the one (whole) has been divided into and the top number tells you how many of these parts you have. The bigger the bottom number, the more parts the one (whole) has been divided into and thus the smaller each part will be. This can confuse those unwise in the ways of fractions, who might connect big numbers with big values. Remember the bottom number tells you how many times the one has been divided up, so $\frac{1}{1000}$, one thousandth, will be a much smaller quantity than $\frac{1}{5}$, one fifth. Those two letters, th, at the end of the number word make a big difference to the meaning of the number word.

For example, the th effect makes one thousandth ($\frac{1}{1000}$), a fraction that is one million times smaller than one thousand (1000).

Now look at the top number. It tells you how many of the fraction parts you have. So in $\frac{7}{10}$, the top number is 7 which means you have 7 out of 10 parts.

In the vocabulary of maths 'out of' means divide.

Multiplying and dividing with fractions

This is a danger zone. But only if you don't know the rules. At the start of this chapter I said that fractions seem to have their own rules. This is even more so when multiplying and dividing with fractions.

Some people rashly (and wrongly) associate multiplying with making answers that are bigger and think that dividing always makes answers smaller. This is not true when using fractions which have a value of less than one.

Let's work from the familiar, in this case a half, to the general and once more build on what you know.

Fractions times whole numbers
Examine the question, 'What is half of £10?' The word 'of' means multiply, so the question needs you to multiply £10 by a half, yet the answer is £5 which is smaller than £10. So multiplying can make numbers smaller.

The explanation is quite straightforward. Again you need to use the maths code. The question can be written as

$$\frac{1}{2} \times £10 = £5$$

What is hidden in this question is the divide sign in the fraction. ½ can also be written as 1 ÷ 2, so the question hides an extra maths operation and is

$$1 \times 10 \div 2$$

In a question like 'What is ¾ of £40?' a knowledge of the maths code is needed again. The question first becomes

$$\frac{3}{4} \times £40$$

The divide sign hidden in the fraction should be used

$$£40 \div 4 = £10$$

£10 is a quarter, one of 4 parts. As we have ¾, then the answer will be $3 \times £10 = £30$

Dividing by fractions

I have just explained how multiplying by fractions can make an answer smaller. It should not, therefore, be too much of a shock to learn that dividing by fractions can make an answer bigger.

Dividing by a fraction whose value is less than one makes the answer bigger.

Let's start with two examples which should be familiar.

1) There are two quarter hours in each half hour. Written as numbers, this statement looks like this;

$$\frac{1}{2} \div \frac{1}{4} = 2$$

2) There are two half hours in every hour. Written as numbers this statement looks like this;

$$1 \div \frac{1}{2} = 2$$

Quite often people give the answer 5 to the question, 'What is 10 divided by a half?' The answer is 20. This makes more sense if the question is re-worded as, 'How many halves are there in 10?' Re-wording is often a good strategy. It has the additional benefit of stopping you from rushing at the question.

To summarise;

MULTIPLYING
by 1 gives an answer which is the same value
by numbers bigger than 1 gives bigger answers
by numbers less than 1 gives smaller answers

DIVIDING
by 1 gives an answer which is the same value
by numbers bigger than 1 gives smaller answers
by numbers smaller than 1 gives bigger answers.

(If you need to know more about fractions, try my book, 'What to do when you can't do fractions, decimals and percentages')

8 Probability

This is a version of fractions, decimals and percentages used, among others, by gamblers and insurance actuaries. It puts the possibility of any event or occurrence on a scale of 0 to 1. A probability of 0 is that the occurrence will be impossible. A probability of 1 is that the occurrence will be certain. Not surprisingly, therefore, a probability of ½ (or 0.5) is a 'fifty-fifty' or evens chance, the most common example of which is tossing a coin and choosing heads or tails. Probabilities can also be expressed as fractions or percentages.

The probability line:

Impossible	Evens	Certain
impossible	evens	certain
0	½	1
0	0.5	1
0	50%	100%

The chance of winning the Lottery with one entry is around 14 million to 1 or ¹⁄₁₄ ₀₀₀ ₀₀₀, which is a number very close to 0! (But it is not 0 because people do win).

When probabilities are given as percentages, the usual 'translation' applies, so a probability of 1 becomes 100% and ½ becomes 50%. Weather forecasters use percentages to indicate the relative probability of rain or snow. A 90% chance of rain means that rain is very likely. A 10% chance of snow means there will be little chance of building a snowman.

Spinning a coin
Probabilities are about the relationship between a successful outcome and all the possible outcomes. To help understand this, think about spinning a coin and calling 'Heads'.

A coin has two sides. If it is a 'fair' coin, the chance of getting a 'head' or a 'tail' will be the same. This means that there are two possible outcomes altogether. If the coins lands as 'heads', this would be the successful outcome. So you have 1 successful outcome. The number of all possible outcomes is 2 and the chance of getting a 'head' on each spin of the coin is 1 out of 2 or ½.

I can write a generalised probability fraction as;

$$\text{Probability} = \frac{\text{the number of successful outcomes}}{\text{the number of all the outcomes}}$$

Let's apply this to throwing a six sided die (die is singular for dice). Say you want to get a four. That means you have 1 successful outcome, a four. The die has six sides and six numbers, so there are six possible ways that the die can land. So all the possibilities are 6.

The probability of throwing a four =

$$\frac{\text{The number of successful outcomes}}{\text{The number of all possible outcomes}} = \frac{1}{6}$$

Now let's look at the probabilities in the stars and horoscopes. . . .

I like reading my horoscope, but there is a probability barrier preventing me from taking it too seriously. Let me explain.

Say the population of England is fortyeight million (an underestimate, but it's going to divide better). There are twelve star signs. Say the birthdays of these fortyeight million people are evenly distributed through the year, then there are about 4 000 000, four million, people sharing each star sign.

I figure three probability related things from this;

It is a certainty that all predictions for a star sign will be vague!

It is probable that at least one of these four million people with the same star sign will fulfil one of these vague predictions ('exciting news about money will land on your doormat').

I think it is a low probability that descriptions of typical Taureans (or whichever star sign) will fit all the four million Taureans in England (except, of course, the bits that I, with total objectivity, agree with).

9   Decimals

All the numbers below are decimals.

   12.5   66.9   0.95   3.25   and even   £25.95

Decimal fractions are another way of writing numbers less than one. In each of the examples above the figures written after the decimal point (the dot) represent numbers less than one, for example the .5 in 12.5 represents a half. The identifying characteristic is the point, called the 'decimal point'. This decimal point separates the whole numbers from the part numbers. It is written immediately after the unit digit.

The unit digit is the focus, the centre of symmetry in this concept.

For example,   28.756

28	.	756
whole numbers	decimal point	part numbers

£54.65

£54	.	65
whole pounds	decimal point	pence (parts of pounds)

Although decimal fractions are like ordinary fractions in representing numbers less than one, they are restricted to tenths (1/10), hundredths (1/100), thousandths (1/1000) etc. Like whole numbers (which work on tens, hundreds, thousands, etc), decimal fractions use place value (see page 13). The place that a digit holds in the decimal fraction influences its value. Basically, the place of a digit in the number tells you whether the fraction is tenths, hundredths, thousandths and so on.

The order is logical. Decimal fractions become ten times smaller each one move right from the unit digit. Just as whole numbers do. The decimal point tells you that you are moving to values less than one.

3	2	8	.	7	5
three	two	eight	decimal	seven	five
hundreds	tens	units	point	tenths	hundredths

There is a natural tendency to focus on the decimal point. It is important as it does tell us that decimal numbers are present. It is not, however, the centre of symmetry. That honour belongs to the unit digit. As the digits go up in place value from the units, they go to tens, hundreds, thousands and so on. As they go down in place value from the units, they go to tenths, hundredths, thousandths and so on.

Money

Money is an everyday example of decimals. Sometimes a dash is used instead of the decimal point, but the principle is the same. The pound is the unit, one whole. Pence are hundredths (since 100 pence are the same as £1) and each ten pence is a tenth (since 10 ten pence coins are the same as £1).

A price is a good example of decimal fractions, though most people do not analyse money in this way. Take £23.95 for example, and look at the meaning of each digit:

£	2	3	.	9	5
two	tens	three units	decimal point	nine tenths	five hundredths

The decimal part of £23.95, that is, .95 is shown as 9 ten pence coins and 5 one pence coins. It could also be thought of using just one pence coins, which are each 1/100 of £1. So .95 could also be shown as 95 one pence coins (which would be quite heavy in your pocket!)

The .95 is $^{95}/_{100}$ of £1   or $^9/_{10}$ plus $^5/_{100}$ of £1.

Money is useful to show the connection between key value fractions and key value decimals.

fraction	money	decimal
1	£1 = £1.00	1.00
½	50p = £0.50	0.50
¼	25p = £0.25	0.25
⅕	20p = £0.20	0.20
$^1/_{10}$	10p = £0.10	0.10
$^1/_{100}$	1p = £0.01	0.01

Counting in decimals

It can help your understanding of decimals if you do a little carefully selected counting in decimals.

Start at 0.1, 0.2, 0.3, 0.4, 0.5, 0.6, 0.7, 0.8, 0.9 . . . what is the next number?

The next number is 1.0 (not 'zero point ten' 0.10). You have moved into the next place value, the units, which is what happens after you reach nine in a place value. In this case you started in the tenths. When you reach 9 tenths, the next number will be in the next bigger place value, that is units.

This is what happens when you count using ten pence coins.

Now start at 0.01, 00.2, 0.03, 0.04, 0.05, 0.06, 0.07, 0.08, 0.09 . . . what is the next number?

The next number is 0.10. You are counting in hundredths. When you reach 9 hundredths, the next number will be in the next place value, that is tenths.

This is what happens when you count using one pence coins.

These are further examples of place value and another application of the same rules.

Adding and subtracting decimal numbers

The first step with adding or subtracting decimal numbers is to line up the unit digits. Sometimes you can check this seeing if the decimal points are lined up, for example with a sum such as;

$$12.3 + 7 + 4.35 + 3.219$$

the lining up gives

```
 12.3
  7
  4.35
  3.219
26.869
```

As with whole number additions, the rule is to line up numbers with the same place values. In the example chosen, the 3 in '12.3' is 3 tenths, the 2 in '3.219' is 2 tenths and the 3 in '4.35' is 3 tenths, so these are all lined up to make sure you add like to like.

This also applies to money, for example, £12 plus £3.50 is added as

```
    £12.00
 + £  3.50
   £15.50
```

The most common mistake when adding decimal numbers is to line them up as though they were whole numbers so 15.4 + 3 is added as

```
    15.4
      3
    15.7    which is WRONG ✗
```

As ever, you need to pause before starting and think what the rules are. In this case, as with all addition and subtraction, once again you have to line up the same place values.

Shopping

The favourite decimals used in prices in shops are 0.90, 0.99 and 0.95. Think how many times you have seen prices like £4.99, £19.95 and sale tickets which say, 'Under £100' and then give the price as £99.99.

It really is a great trick, and a fairly international one at that. Shoppers see £14.99 and focus on it as 'fourteen pounds'. The 99 pence is as much as the shop can add to £14 before it moves to £15. It has the benefit of shoppers seeing £14, yet paying as close as is possible to £15!

When you get to big items like cars and houses, this place value subterfuge is still used. Cars at £9 950 and £12 995. Houses at £139 950 and £349 500. A little knowledge of place value in numbers combined with realistic cynicism can be useful.

I feel that this pricing device is aimed, though I think often unconsciously, at inchworms (see Chapter 1) who tend not to scan along a number, but focus on the first digits. Also they see numbers exactly as written rather than rounded up or down to some more convenient value. I suspect grasshoppers are immune to this pricing strategy.

10 Measuring

Measuring length in metric units

I am sure that most people would rather avoid fractions and decimals. This is one of the reasons we have "pence" rather than "one hundredths of a pound".

The metric system allows us to avoid decimals by using a prefix instead of a decimal point.

If £1 is the basic unit of money then 1 metre is now the basic unit of length. The metre is too long for some measurements, so we use prefixes, as in "*milli*metre" as a way of dealing with fractions of a metre.

The metre is also too short for big measurements so there are prefixes to cope with this, too, for example a *kilo*metre, but more of that elsewhere.

The metric system is based on ten and powers of ten (10, 100, 1000, etc). The metric fraction prefixes used are, of course, also based on ten ($\frac{1}{10}$, $\frac{1}{100}$, $\frac{1}{1000}$, etc).

The metric system uses prefixes to avoid using a decimal point. They are:

deci, d (not used a lot),
centi, c
milli, m (used a lot).

The table shows how this is done for metres.

$\frac{1}{10}$ metre 0.1 metre 1 decimetre 1dm $\frac{1}{10}$ m

$\frac{1}{100}$ metre 0.01 metre 1 centimetre 1cm $\frac{1}{100}$ m

$\frac{1}{1000}$ metre 0.001 metre 1 millimetre 1mm $\frac{1}{1000}$ m

The prefixes are an alternate way of writing a fraction or a decimal. Instead of writing $\frac{59}{100}$ metre or 0.59 metre, we write 59 cm or 590 mm. Instead of writing $\frac{734}{1000}$ metre, or 0.734 metre, we write 734 mm.

Builders merchants and DIY stores are more likely to use mm rather than cm for fractions of a metre, for example;

20mm is $\frac{20}{1000}$ metre or 2cm

100mm is $\frac{100}{1000}$ metre, which is $\frac{1}{10}$ metre or 10 cm

1500mm is $\frac{1500}{1000}$ metre, which is 1.5 or $1\frac{1}{2}$ metre.

Big distances

A metre is approximately one long stride at full stretch, but it is still not a big distance in terms of, say travelling from London to York. For these distances we use the *kilo*metre. The prefix, kilo, which has been added to metre, means "one thousand times more". A kilometre is 1000m.

Measuring length in imperial units

The USA and sometimes the UK still use imperial units. The relationship between the Imperial units is based on a variety of numbers including 12 and 3.

12 inches = 1 foot	1 inch can be written as 1 in or 1"
3 feet = 1 yard	1 foot can be written as 1ft or 1'
36 inches = 1 yard	1 yard can be written as 1yd.
1760 yards = 1 mile	
5280 feet = 1 mile	

Conversions from metric to imperial lengths and vice versa are shown in the table below:

accurate	estimate
1 in = 2.54 cm	1 in = 2.5 (2½) cm
1 cm = 0.3937 in	1 cm = 0.4 in
1 ft = 30.48 cm	1 ft = 30 cm
1 yd = 0.9144 m	1 yd = 90 cm
1m = 1.0904 yd	1 m = 1.1 yd
1 mile = 1.6093 km	1 mile = 1.6 km or use ⅗ km
1 km = 0.6214 miles	1 km = 0.6 mile or use ⅝ mile

The estimate values provide the easy way to convert between metric and imperial units. They provide key values (a technique which we shall use for foreign currency conversions in Chapter 12).

For example, to convert 240 km into miles:

The answer will be a smaller number (because miles are bigger than kilometres). The key (approximate) values are   10 km = 6 miles
and 8 km = 5 miles

So choose the key value which fits the numbers you have the best.

For 240 km the 8 km = 5 miles is good.

Now develop the key value, using simple multiples, starting with 10×.

       80 km = 50 miles
      160 km = 100 miles
      240 km = 150 miles

This estimation strategy can also be used for mph and kph conversions.

To convert 60 inches into cm:

The answer will be a bigger number.

The key value is developed from the estimate   1 inch = 2.5 cm
10 inches = 25 cm

So   20 inches = 50 cm
      40 inches = 100 cm
      60 inches = 150 cm (which also tells us that 5 feet = 1.5 metres)

To convert 80 cm to inches, the key value is developed from

$\qquad$ 2.5 cm = 1 in

So $\qquad$ 10 cm = 4 in

Now do simple multiples of this key value;

$\qquad$ 20 cm = 8 in
$\qquad$ 40 cm = 16 in
$\qquad$ 80 cm = 32 in

The relationship between yards and metres allows a different technique for conversion. 1 metre = 1.1 yards. So a metre is 10% more or $\frac{1}{10}$ more than a yard.

This means that to convert metres to yards, you simply add 10% or $\frac{1}{10}$.

For example 6 metres in yards is 6 plus $\frac{1}{10}$ of 6, which is 6.6 yards.

Measuring volume in metric units (volume is sometimes called capacity).

The basic unit of volume in the metric system is the litre, shortened to l. A litre is quite a large volume (about 1¾ pints) so again prefixes are used to deal with smaller volumes. The prefixes in the metric system are always the same whatever the unit, so:

$\qquad$ d for deci, which is $\frac{1}{10}$

$\qquad$ c for centi, which is $\frac{1}{100}$

$\qquad$ m for milli, which is $\frac{1}{1000}$

This gives decilitres, dl, centilitres, cl and millilitres, ml.

A can of soft drink is usually 330 ml $\quad \frac{330}{1000}$ litre $\quad$ about ⅓ litre
A bottle of whisky is usually 70 cl $\quad \frac{70}{100}$ litre
A bottle of wine is often 75 cl $\quad \frac{75}{100}$ litre $\quad$ ¾ litre
A tin of gloss paint is 500 ml $\quad \frac{500}{1000}$ litre $\quad$ ½ litre
A pint of milk is 568 ml $\quad \frac{568}{1000}$ litre $\quad$ just over ½ litre.

I have yet to spot the rational for producers using centilitres for some liquids and millilitres for others

Measuring weight (mass) in metric units

The basic unit of weight in the metric system is a little different in that it already has a prefix, the *kilo*gram, kg. The prefix means kilo or 1000 times. This means that the kilogram is 1000 grams. (A gram weighs about the same as a drawing pin). So, for weights less than a kilogram instead of a prefix the gram, g is used. One gram is $\frac{1}{1000}$ of a kilogram.

A bag of sugar weighs 1 kg

A packet of butter weighs 250 g $\frac{250}{1000}$ kg $\frac{1}{4}$ kg

A packet of fudge weighs 100 g $\frac{100}{1000}$ kg $\frac{1}{10}$ kg

Conversion of kilograms to pounds uses the same strategy we used for converting metres to yards. 1 kilogram is about 2.2 pounds.) .2 is one tenth ($\frac{1}{10}$) of 2. So to convert kilograms to pounds, multiply by 2 and then add one tenth of that.

For example, 80 kg to pounds

First step	$80 \times 2 = 160$
Second step	$160 \div 10 = 16$
Third step	$160 + 16 = 176$ lb

Big weights

Once again weight (or mass) is a little different to other metric units. A thousand kilograms should be a Megagram, because the pre-fix Mega means 1 000 000, one million, which is one thousand thousands (hence a Megastar means someone pretty famous).

In this rather individual situation for weight, one thousand kilograms is called a 'tonne' (not a ton, which is the Imperial equivalent, 2240 lb).

1000 kg is 1 tonne

(1 ton and 1 tonne are actually very close in value)

Measuring weight (mass) in imperial units

The imperial units for weight are summarised in the table below

16 ounces (oz)	= 1 pound (lb)
14 pounds	= 1 stone (a UK unit, not used in the USA)
2240 pounds	= 1 ton
1000 kg	= 1 tonne

accurate	estimate
1 ounce (oz) =28.35g	1 ounce = 30g (though 25g is often used)
1 gram = 0.0353 ounces	1 gram = 0.04 ounces (1/25 oz)
1 pound (lb) = 0.4536kg	1 pound = 0.45 kg (just less than ½ kg)
1 kilogram = 2.205 lb	1 kilogram = 2.2 lb

11 Percentages

We see many uses of percentages in 'real-life'. It is one of the important concepts for us to understand, particularly as it is often applied to money. For example:

Wage rises
Value Added Tax (VAT)
Inflation
Bank interest rates
Credit cards
Sale offers in shops and catalogues
Deposit on buying a house or car
Tips in restaurants
Mortgage interest
Statistics
Weather (for example, 'a 50% chance of rain today')

Percentages are another way of dealing with parts of one. Unlike fractions the values they represent are easy to compare, largely because they use the easy concept of the numbers 1 to 100.

Being able to understand and have a feel for the value of percentages are key skills for life.

The word percent means 'out of 100'. Per means 'out of' and cent means '100'. So percentages are really a special, restricted value (name) fraction. All percentages are hundredths $\frac{1}{100}$. We take the symbols '$\frac{1}{100}$' and transform them into '%'. When we say 'percentages' we are talking about fractions that are hundredths, but most people consider 'percentage' much easier to understand than a fraction.

Percentages have a particular advantage. It is quite difficult to compare the values of different fractions, for example deciding which is bigger, $^{15}/_{32}$ or $^{35}/_{76}$. This task is very much easier with percentages. It is obvious which is bigger out of 47.9% and 46.1% (which means that $^{15}/_{32}$ is bigger than $^{35}/_{76}$) even though the values are close.

The ease with which percentages can be compared is useful when comparing interest rates on different savings accounts or mortgages or interest rates on different credit cards. Weather forecasters sometimes use the easy comparability of percentages as a clear way to indicate how likely we are to get rain. A chance of rain of 80% suggests that it would be wise to take an umbrella when you go out. A chance of rain of 10% suggests you might risk leaving the coat at home. (I always get that 10% of rain falling on me just as I go to the shop, post, train or whatever!)

So, let's get the big picture. . . .

100% is the whole. It is one. It is all of something.

This is used as an everyday saying. If someone asks you for 100% effort, they want all of your effort (unlike football managers who seem to want 110% or more from their players).

So anything less than 100% is less than all of it!

It follows, then, that we can set up some key values for estimations and for building other values:

50% is one half, $\frac{1}{2}$.
25% is one quarter, $\frac{1}{4}$.
33$\frac{1}{3}$% is one third, $\frac{1}{3}$.
10% is one tenth, $\frac{1}{10}$.
1% is one hundredth, $\frac{1}{100}$.

Once again these key values can be inter-related so that easy value numbers can be extended to give access to harder numbers.

An example will show how these key percentage values can work;

Our local travel agent adds 2% to your holiday bill if you pay by credit card. What is the extra payable on a holiday costing £750?

1% is $\frac{1}{100}$, so to work out 1%, divide £750 by 100.

£750 ÷ 100 = £7.50

A 2% surcharge will be twice as much,   2 × £7.50

The 2% surcharge is £15.00

Another way to approach this percentage calculation is to interpret 1% as meaning 1 in every 100 (so 20% is 20 in every 100, 25% is 25 in every 100 and so on). This is an early mention of a ratio / proportion type calculation (see chapter 12).

For the travel bill percentage, 1% of £750 is £1 in every £100. £750 is 7 and ½ hundreds, so a 1% surcharge must be seven and a half pounds, £7.50. By proportion, 2% must be twice as much surcharge. Twice £7.50 is £15.

You can use the key values of 1%, 10%, 50% to get other values and thus get close approximations for many other percent values. Once again you make use of the relationships between numbers, for example;

5% is half of 10%

So if you can work out 10% of a number, if you then half that 10%, you have the 5% value, as in

£800 . . 10% of £800 is one tenth of 800, that is 800 divided by 10

£800 ÷ 10 = £80.   £80 is 10% of £800, so 5% of £800 must, by proportion, be a half of £80, that is £80 ÷ 2 = £40.

Another example:

12.5% is a half of 25% and 25% is a half of 50%. So, 12.5% is a ¼ of 50%

If you can work out 50% of a number, you can halve it to get 25% and then halve it again to get 12.5%. For example:

£104 . . . 50% of £104 is half of £104, that is £104 divided by 2

£104 ÷ 2 = £52   £52 is 50% of £104

25% of £104 must be half of £52

£52 ÷ 2 = £26

And 12.5% is calculated as half of £26

£26 ÷ 2 = £13

(To work out 12.5% you divide the 100% value by 2 three times. In other words you halve, halve and halve again. You could also do this in one step if you can divide by eight).

If you want to work out 20% you can double the 10% value. This is a way to calculate the current rate of VAT in the UK.

For example, estimate the VAT payable on a computer costing £950. VAT is 20%.

10% of £950 is one tenth of £950, which is £950 divided by10
£950 ÷ 10 = £95 . . . so 10% of £950 is £95
So 20% of £950 is 2 × £95 = £190

The VAT payable on a £950 computer is £190

Do not think that percentages are restricted to values of 100% or less. It is possible to have values over 100%. For example 200% is twice the value of the original value.

Once you have grasped these basic and key percentage values, you can then progress to combining them to make even more percentage values easy to calculate.

For example you can add 10% and 5% to get 15%
or you can add 25% and 50% to get 75%
or you can add 1% and 10% to get 11%

The easy, key values open up a wide range of percentages to quick calculations, which are useful for checks on calculator results, or as estimates for tips, or VAT additions to bills and shopping prices.

Once again the maths is about taking the 'easy' numbers and using them in different topics. So many topics are closely connected and are merely extensions of the same idea.

For example;

What is the VAT payable on a builder's bill for £480? (In the UK in 2011 the VAT is 20%)

Work out 10%, then double this to get 20%

> 10% is £480 ÷ 10 = £48
> 20% is £48 × 2 = £96

Have you ever thought about the percentage interest rates charged by some stores for hire purchase? Sometimes these hover around the 30% level, which is close to ⅓. So a television, costing £600 would attract about £200 interest in one year (if you paid nothing off).

10%. A special case and a reminder of a key skill.

It is a recurring theme in this book that maths ideas keep reappearing, often in different disguises. 10% is a good example of this. The quick way to calculate 10% of something is to divide by 10. Being able to divide by 10 (and by 100 and by 1000) should be a basic and essential skill.

Reminding you how to divide by 10. . . .

When you divide a number by one it stays the same value and (with the exception of a special case used in fractions) looks the same.

10 is one ten. When you divide a number by ten (one ten) it looks virtually the same, but there is an important difference, the digits in the result are the same and in the same order, but they now have different place values. For example,

> 1 234 567 890 ÷ 10 = 123 456 789

Each digit now has a place value worth $\frac{1}{10}$ of its previous value (for example, the 7 has moved from being 7000 to become 700). This makes sense, since if you divide a number by 10, then all parts of the number should get 10 times smaller!

This last example had a convenient 0 in the units place. If another digit was in the units place, then dividing by 10 takes the answer into decimals, for example

$$98\ 765 \div 10 = 9\ 876.5$$

Again the digits in the answer are the same and in the same order as the original number and, again, each digit in the answer now has a place value which is $\frac{1}{10}$ of its previous value. For example, the 9 was 90 000 and has become 9 000 and 5 units have become .5 (5 tenths). This is a good way to start the calculation. Focus on one digit and make sure its place value goes down to $\frac{1}{10}$ of its previous place value. Then make sure that all the other digits are in the same order as the original number. This will make all these other digits also $\frac{1}{10}$ of their original value. AND don't forget you will find the unit digits becoming decimal digits.

The same principle applies to dividing by 100, one hundred. This will take digits down two place values (there is a reminder built in to the numbers here, 10 has one 0 and the digits move one place, 100 has two 0's and the digits move two places). This two place move makes sense if you start with a simple example, say $400 \div 100 = 4$. The 4 has moved down two place values.

In a harder example, such as

$$456\ 700 \div 100 = 4567$$

the same two place value move has occurred. You can check by focussing on the hundreds digit, 7, which has become a units digit in the answer.

If the number to be divided doesn't have a convenient supply of zeros at the end, you will have an answer which moves into decimal place values.

$$456\ 789 \div 100 = 4567.89$$

Again if you focus on that hundreds digit, the 7, it has moved to the units place value.

This pattern and these strategies apply to any division further in the 10's series, 1000, 10 000, 100 000, 1 000 000 and so on. To check, focus on an obvious digit whenever possible. For example when dividing by 1000's check that the thousands digit moves down to the units place. (You can use a pattern here. 1000 has three zeros, so the digits have to move three places).

Other percentage calculations.

These need a calculator and are for when you want to obtain a precise answer rather than an estimate.

Type 1. These are questions which ask you to calculate an exact percentage of a number. For example;

> What is 35.8% of £660?

First you have to read and interpret the formula. You need to remember the maths meaning of the word 'of' as in '35.8% **of** £660.'

> 'Of' means 'times' or 'multiply'.

% means divide by 100 (although the % key on a calculator does both divide by 100 and equals)

So, if you use a calculator, the sequence is;

1) key in 35.8
2) key in x
3) key in 660
4) key in %

The sequence follows the maths equation   35.8% of £660   and gives an answer of £236.28. The percent (%) key divides by 100 *and also does the equal (=) key so that the answer comes straight up on the screen.* It does two keys for the price of one!

Remember to check your answer against an estimate. In this example, you could use 50%. This gives £330. Since 35.8 is less than 50 you would expect the accurate answer to be 'quite a bit' less than £330.

Another example of percentages, this one about our 20% rate of VAT.

A builder's estimate for repairing a roof is £1220 + VAT. What is the total amount to pay?

There are two options here. One is to work out 20% (VAT rate) of £1220 and add this to £1220. The other is to work out $100 + 20 = 120\%$ of £1220, which will take you straight to the total bill.

Let's try the second option;

Rephrase the question. What is 120 % of £1220? The 'of' means 'multiply' and '%' means 'divide by 100'.

The formula is   120% of £1220

On a calculator, the sequence of keys uses the % button, which automatically divides by 100 and does the = button, too.

1) key in 120
2) key in x
3) key in 1220
4) key in % (which also operates the = key)

Answer £1464

The same strategy can be used with discounts. For example;

A car dealer offers a special summer discount of 12% on a second hand car priced at £6495. How much is the discount price?

A 12% discount means the new price will be $(100 - 12)\%$, that is, 88% of the original price. The new price is 88% of £6495, so get the calculator and look out for an answer about 10% less than £6495 (about £700 less).

1) key in 88
2) key in × (for the 'of')
3) key in 6495
4) key in %

> The answer is £5715.60

Questions which ask you to calculate a percentage from two numbers.

This type of calculation is useful for comparing data.

For example;

> A pupil scores 47 marks in an examination where full marks are 80. What is her percent score?

Often with a word problem (on any topic), it is a good idea to read and then reword the question until it takes a form that makes sense to you. For this question, you could rephrase the pupil's score to be, '47 out of 80.' The words 'out of' make the question a fraction $^{47}/_{80}$, which means 47 divided by 80. To make the fraction a percent, multiply by 100;

$$^{47}/_{80} \times 100 = 58.75\%$$

Alternatively, once you have decided this is a 47/80 question and thus identified the question as a division, you can use the % key on the calculator. There are four steps;

1) key in 47
2) key in ÷
3) key in 80
4) key in %.

(This time the % button automatically multiplies by 100 and operates the '=' key).

The sequence of keys in these percentage calculations follows the formula.

12 Proportion / Ratio

Proportion (or ratio) is basically a division process, but with an extra step.

Division is about sharing equally. Proportion is about shares that are not all equal.

As a student I was forced by lack of finance, extreme hunger and flat mates who were the culinary equivalent of illiterate, to learn how to cook for myself. For the first time I appreciated the practical uses of proportion.

Later in my life I became a (mortgaged) house owner. Again lack of finance, crumbling walls, muddy paths and friends who were the DIY equivalent of illiterate forced me to learn how to mix (and use) mortar and concrete. For the second time I appreciated the use of proportion.

Both these real life examples of proportion take a rough (approximate) attitude to complete accuracy. But the principle you use is the same as with accurate proportion calculations.

Start with food. A simple recipe for a crumble topping is;

    150g flour
     75g butter
     75g sugar

150g is twice 75 g, so the proportions (or ratio) are 2 parts of flour, 1 part of butter, 1 part of sugar, which makes a total of 4 parts (and 150g + 75g + 75g = 300g of crumble topping). Once I know that ratio I can make any quantity of crumble. The approximate/estimate aspect comes from using, say a spoon or cup to measure out the ingredients. 2 cups of flour. 1 cup of butter. 1 cup of sugar.

When dealing with proportion (or ratios) there are two numerical areas on which to focus;

the *individual* proportional parts;   in this example, 2, 1 and 1

the *total* number of proportional parts;   in this example
2 + 1 + 1 = 4

Although the recipe I used worked with 75g, 75g and 150g, I could
use any values in the 1 to 1 to 2 proportions, say 1kg of sugar, 1kg
of butter and 2kg of flour. This makes a rather large amount of
crumble topping, but it will be in the right proportions. The total
weight is 1 kg + 1kg + 2kg = 4kg.

Now, for the concrete experience of proportion (not to be confused
with crumble mix, but similar in some respects):

The proportions (ratio) I used for mixing my concrete were by
volume not weight. I interpreted 'one volume part' as a heaped
amount on a shovel, so for a dry mix of concrete the proportion
parts are;

1 cement
2 sand
4 aggregate (chippings)

The total number of parts is 1 + 2 + 4 = 7.

So if it takes 35 heaped shovels to almost fill my wheelbarrow, I
will need 35 ÷ 7 = 5 times each proportion part;

5 × 1 = 5 shovels of cement
5 × 2 = 10 shovels of sand
5 × 4 = 20 shovels of aggregate

To do this calculation, I needed to know the total number of proportion parts (in this example, 7). I divided the quantity I had to
make (35) by this total number of parts to find out how many lots
of each part was needed. I needed 35 ÷ 7 = 5 lots of each proportion part.

There is another type of situation where you may meet proportion.

In this example, you have to divide something up into proportions. (Ordinary division divides things up into equal parts, but
proportion parts are unequal. So, to find one unit part, first you
have to add up all the proportion parts before you do the division).

Say a lottery syndicate has five members, Mr Black, Mr White, Mr Orange, Mr Red and Mr Green. Each Saturday, Mr Black pays in £3, Mr White pays in £2, Mr Orange, Mr Red and Mr Green each pay in £1. They agree to share any winnings in the same proportion as their 'investments'. To show how this works out, let's say that they win £4000.

The parts are 3, 2, 1, 1, 1
The total number of parts is $3 + 2 + 1 + 1 + 1 = 8$
£4000 has first to be divided up into 8 parts

£4000 ÷ 8 = £500

So 1 part is £500
2 parts are $2 \times £500 = £1000$
3 parts are $3 \times £500 = £1500$

To summarise (and check):

Mr Black	£1500
Mr White	£1000
Mr Orange	£ 500
Mr Red	£ 500
Mr Green	£ 500
Total	£4000

Three other everyday examples where proportion is used are maps, plans and architectural models. An artistic example would be a drawing (traditional rather than Picasso) of a human body. It would be obvious if, say the legs were out of proportion to the rest of the body (except with Barbie dolls, but then they are not human) or if the eyes were drawn too high on the head.

Inverse proportion

So far we have looked at direct proportion, for example if you were making crumble mix and you doubled the amount of flour, you would have to double the amounts of the other two ingredients as well. In direct proportion if you half one part you half the other parts, if you multiply one part by ten, then you have to multiply all the other parts by ten.

There is also an inverse proportion, where, for example doubling one quantity is balanced by halving the other quantity. For example, if you travel along a motorway for 120 miles at a constant 40 miles per hour, you will travel for 3 hours (and annoy a lot of people). If you then do the same journey at twice the speed, 80 miles per hour, your travelling time will be halved to 1½ hour (and you will be breaking the legal speed limit).

Another real life example, often used in maths books is the one about the mythical bricklayers. . . .

If one bricklayer builds a wall in 12 hours, then, providing they all work at the same rate and keep out of each others' way, three bricklayers will build the same size wall in 4 hours. Using three times the bricklayers gets the job done in one third of the time (in theory, but then much of the 'real life' maths in text books is very much 'in theory').

If you look at this bricklayer example in terms of the numbers involved, you will see that there is a constant value in multiplying the number of bricklayers by the time taken to build the wall.

number of bricklayers × time to build the wall

$$1 \times 12 = 3 \times 4$$

This is an example of the relative size (value) of the two multipliers for a given product, for example, if the product is 24, the multipliers can be;

$$1 \times 24 = 24$$
$$2 \times 12 = 24$$
$$3 \times \ \ 8 = 24$$
$$4 \times \ \ 6 = 24$$
$$6 \times \ \ 4 = 24$$
$$8 \times \ \ 3 = 24$$
$$12 \times \ \ 2 = 24$$
$$24 \times \ \ 1 = 24$$

As the bold numbers get bigger, the plain numbers get correspondingly smaller.

Foreign exchange and proportion

When you travel abroad you have to cope with a different currency with a different value. For example, in March 2011 £1 was equivalent to 1.58 USA dollars.

If you know the rate of exchange, then you can work out the cost of meals, shopping, hotels and so forth. This is vital for keeping to budget.

There are accurate ways to calculate the exchange values, but you will need to have a calculator on hand. There are also effective estimate ways, which are really a variation of proportion. And they can be a lot handier than using a calculator. These methods deal with those occasions when you do not need to know exactly what the exchange money value will be, but you want to be close enough to not overspend on a drink or a meal or a souvenir.

Let's look at the estimate method, using some exchange rates which were current in Spring 2011. Although these examples are specific, I will try to build up a general method for you to use. The strategy uses proportions involving the 'easy' numbers.

In March 2011 the exchange rate for £1 was:

Australia	Dollar	1.54
Eurozone	Euro	1.14
Hong Kong	Dollar	12.16
Norway	Krone	9.00
Switzerland	Franc	1.50
Turkey	New lira	2.47
USA	Dollar	1.58

For example, the exchange rate for the USA dollar is 1.58 dollars to £1. This is fairly close in value to $1.5, which a lower exchange rate, but close enough to give a reasonable estimate. So a shirt priced at $30 is approximately £20. The next move may be to refine this a little. This brings in the generic question, "Is the answer bigger or smaller?"

Let's use two wild estimates to illustrate 'bigger or smaller'.

First let's round **up** $1.58 to $2. The $30 shirt now exchanges to £15, lower than the true price (of £18.99) and the first estimate of £20.

Now round **down** $1.58 to $1. The $30 shirt now exchanges to £30, higher than the true price.

So, rounding **up** an exchange rate might make you think you are spending **less** than you really are.

Rounding **down** may make you think you are spending **more** than the real price.

To illustrate further, take the Norwegian Krone, which is currently 9 to £1.

If the item was priced at 63 Krone, then the true exchange rate of 9, gives a sterling price of £7.00. If you round up the exchange rate to 10 Krone, then the estimated price is £6.30 which is lower than the true price. **Rounding up** the exchange rate gives estimates which are **below** the true price.

A second example . . . the exchange rate for the Eurozone is 1.14 euros for £1. This is fairly close to a 1 to 1 proportion. So, in terms of estimating, that may be enough to give the first answer, for example, £20 is close to €20. Back to the question, "Is the answer bigger or smaller?" You have rounded down, so the estimated price is higher than the true price.

For general pragmatic purposes, an estimate may be enough. It may stop you over-spending or give you an exchange that is close enough for holiday purposes. In which case you could set up some key exchange values. Back to the USA dollar as an example.

The exchange rate for US dollars at the time of writing this book is £1 for $1.58. It can help if you can find easy numbers for a reference exchange rate.

The easy number is $1.5. This is a lower rate, so the £ estimates will be higher than the true value estimates. Set up the key values first

US $	15	30	150
UK £	10	20	100

Now use proportion to fill in more values. For example, if you divide $15 by 3 to get to $5, then you divide £10 to get £3.33

US $	5	15	30	45	50	100	150
UK £	3.33	10	20	30	33	66	100

This could be seen as two number lines merged together:

US $ 0	30	50	100	150
UK £ 0	20	33	66	100

With the table or with the line, the in-between values can be estimated with good accuracy. The line is about seeing proportions visually, the table is about seeing proportions numerically. For example, £50 is half of £100 and this gives $66 halved to be $33.

All you need in your pocket is a small card with the key values plus some derived values and you can look like a mental arithmetic foreign exchange wizard.

I think we are conditioned to believe that we should be able to do detailed and accurate calculations completely in our heads and that charts like these are soft options. We should learn to do what is effective for us. I would argue that for many people these little charts are more effective than calculators for quick estimations of currency values. They certainly provide sensible support for mental calculations.

13 Averages

Everyone should have some idea about some statistics. This does not mean you have to be able to do complex statistical calculations (actually a good Spreadsheet will do that extremely quickly), but you should be able to take a wary and almost cynical look at what is too often presented as persuasive statistics.

Averages are often used in statistics. They are a good place to start to hone our wary and cynical appraisal.

First, what are averages and how are they calculated?

An average is usually taken to mean something in the middle or something typical, like an average family. (I selected this example to sow the first seeds of caution in your mind).

There are, in fact, three averages in common use. The most common of these is the arithmetic average (also called the mean value). To calculate the arithmetic average you add up all the values and then divide by the number of values you added up.

This can be written as a formula;

$$\text{Arithmetic average} = \frac{\text{The sum of all values}}{\text{The number of values added}}$$

An example;

In one week in February, the temperatures each day were 12°C, 10°C, 9°C, 6°C, 4°C, 3°C and 5°C. What is the average temperature for the week?

The sum of all the temperatures is:
$$12 + 10 + 9 + 6 + 4 + 3 + 5 = 49°C$$

The number of values added is 7

Average temperature = $49 \div 7 = 7°C$

This is quite a sensible average. All the values were reasonably similar, so the average was a middle example and with not too much of a spread of values.

This spread is called the *range*, calculated by subtracting the lowest value from the highest value, (so $12 - 3 = 9°C$ here).

Averages are not always as typical or representative as they should be, which is why the range can help us judge the worth of the average.

The average salaries of five men in a factory are £15 000, £17 000, £16 000, £20 000 and £200 000. What is the average salary at the factory?

The sum of all the salaries is;

£15 000 + £17 000 + £16 000 + £20 000 + £200 000
$$= £268\ 000$$

The number of salaries is 5

The average salary is £268 000 ÷ 5 = £ 53 600

The average has been distorted by the one large salary. If the range of values was quoted as well as the average, this would help to make the picture a little clearer. The range is £200 000 – £15 000 = £185 000.

Another less than typical average value is average speed;

A family set off in their car to make a 120 mile journey, mostly along the A 38. The journey takes 3 hours. What is the average speed?

Average speed is calculated by dividing the distance travelled by the time taken

$$\text{average speed} = \frac{\text{distance travelled}}{\text{time taken to travel}}$$

Put into the equation (formula) the values from the question;

$$\text{average speed} = \frac{120 \text{ miles}}{3 \text{ hours}} = 40 \text{ mph}$$

If you think about a car journey on non-motorway roads, the speed is not constant, the cars stop for traffic lights, slow up behind tractors. Children may need to stop for toilets or a drink. The average speed gives only very basic information about the journey.

Treat averages cautiously. They can only tell you some of the information needed towards the whole picture. Look at how much data was used to create the average. Some adverts make claims based on very small samples.

50% is used to indicate the middle value, the average. I often think education officials will only be satisfied with teachers when all children are above average!

14 Angles

The most familiar angle is 90°. It is in such common use that it has a name, the *right angle*, as well as its number value.

Unfortunately key angles do not use the easiest numbers. As I have just said, a right angle is 90° (when 100° would have been easier to work with numerically). A complete circle is 360°.

Some peoples' understanding of angles is blurred because they do not realise that when two lines meet at an angle, the angle size is not dependent on the length of either line (see figure). The length of the meeting lines does not affect the value of any angle.

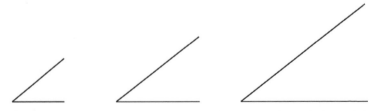

A right angle is a quarter of a circle. If you turn through four right angles (4 × 90° = 360°), then you have turned through 360°, which is complete circle.

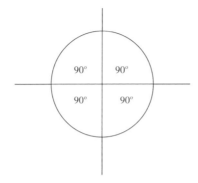

The U turn, immortalised by politicians, is a turn through 180°, or two right angles.

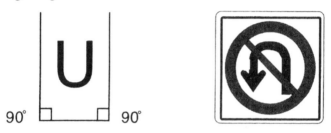

The right angle provides your key reference for other angles. As ever, the idea is to provide some key values to act as references, comparisons and estimations. Halving is always a good start. . . . For example, 45° is half of 90°

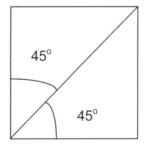

A clock can help you picture 30°. The circular face of a clock means that the second, minute and hour hand all turn through 360° every time they go round the clock once.

If you focus on the 12 hours around the clock, then the hour hand turns through 90° from 12 to 3, and again from 3 to 6, from 6 to 9 and from 9 to 12 (and any other consecutive 3 hour time). If we pick on the 12 to 3 move, which is one quarter of the clock, it is obvious that the angle turned is 90°, one quarter of a complete turn.

This must mean that the hour hand turns through one third of 90° every time it moves from one hour to the next. For example, when the hour hand turns from 2 to 3, it moves through 30° (one third of 90°)

There is a 30° degree angle between every consecutive hour digits.

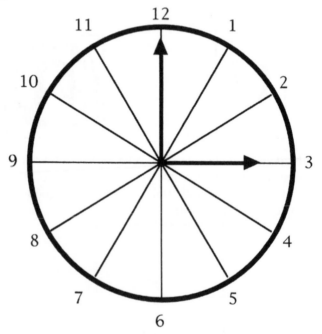

Triangles

The three angles inside a triangle always add up to 180°

A triangle has three angles. The three angles inside any triangle add up to make 180°, no matter what the separate angle values may be. Check this out by cutting up a triangle and putting the three angles together as shown. It will always result in 180° (or two right angles).

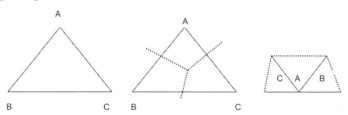

Teaching has taught me that certainty is a bold and often risky stance. I think that is why statements like 'the three angles in any triangle add up to make 180° 'fascinate me with their power of their certainty. That's the triangle rule. No exceptions.

Some shapes

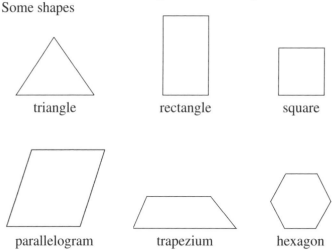

triangle   rectangle   square

parallelogram   trapezium   hexagon

15 Time and Clocks

Time in everyday use attracts two attitudes, precision and approximation. If you have to catch a coach you have to be ready at the bus stop at the right time. If you are walking your dog you are unlikely to say, "I'll be back in 33 minutes and 47 seconds." It's more likely to be, "I'll be about a half hour." The demands for a precise use of time are less frequent in everyday life, so we have less practice and we tend, therefore, to be less skilled at this accurate aspect of time.

Practice is good if we practise the right things. Better still if we understand what we are practising. Our brains are designed to forget as well as to remember. By practising we top up the memory.

Although time uses the numbers 0, 1, 2, 3, 4, 5, 6, 7, 8, 9, 10, 11 and 12 it does have two differences to normal counting. One is that the numbers on an analogue clock are arranged in a circle (rather than a line). This means that the cycle of numbers repeats every 12 (or 24 hours when using the 24 hour clock). This repeating also applies to minutes and seconds, which are based on 60.

So, there are 12 hours round the clock face for the hour hand, 60 minutes around the clock face for the minute hand and 60 seconds around the clock face for the second hand. The clock face acts as three circular number lines (though the numbers for the second and minute hands are the same).

When you count round a clock in hours, the sequence is 1, 2, 3, 4, 5, 6, 7, 8, 9, 10, 11, 12. Then, providing you are not working on a 24 hour clock, it all starts again at 1. The same applies to minutes and seconds, when you get to 60 minutes or 60 seconds, you start over again at 1.

This leads to a modification compared to the way you add ordinary numbers when adding seconds, minutes and hours. The modification requires an awareness of the repetitive number patterns for time and the appropriate adjustment to these 12, 24 and 60 number patterns.

Some examples;

a) 40 minutes plus 30 minutes = 70 minutes. This is greater than 60 minutes, that is, more than 1 hour. 70 minutes is 60 minutes and 10 minutes, that is, 1 hour and 10 minutes.

So 40 min + 30 min = 1 hr 10 min

b) A 4 hour journey starts at 10 pm. What time does it finish? This can be done as a two stage problem, using 4 as 2 + 2. 10 + 2 takes you to 12 midnight. 12 midnight + 2 takes you to 2 am.

So the addition looks like you have added wrongly. In this example, it looks as though 10 + 4 = 2, which, of course would be nonsense if the numbers being used were not time numbers.

If we did this calculation using the 24 hour clock, then 10pm becomes 22.00.

22.00 + 2 = 24.00 (midnight)

The remaining 2 hours take us to 02.00

c) A flight leaves at 20.00 and lands 8 hours later. What is its arrival time?

$$20.00 + 8 = 04.00$$

This can also be done in two stages, going to 24.00 hours first;

$$20.00 + 4.00 = 24.00, \text{ then}$$
$$24.00 + 4.00 = 04.00$$

Again, if you just looked at the raw addition, without knowing that it was about the 24 hour clock, you might think someone was fooling around with maths. In the previous example you could see that 10 add 4 does not normally equal 2. The same is true of the second example, 20 plus 8 does not normally equal 4. For number work with time you have to be very aware of the role of 12, 24 and 60 as "start again" numbers.

Example (c) was an example of an additional complication for work with time . . . the twenty-four hour clock . . . so beloved by the people who construct train and plane timetables and thus is used by travel agents.

There are 24 hours in a day, but, under normal circumstances we work on two circuits of a 12 hour clock and use the letters am to indicate morning (the first twelve) or use pm to indicate afternoon / evening (the second twelve). So 7 am is a morning time and 7 pm is an evening time. (The first is a breakfast time and the second is a cinema time).

In theory the 24 hour clock should remove any confusion over the time of an aircraft flight or a train time. In theory!

Most people use the 12 hour system far more often than the 24 hour system, so the 24 hour system is far less familiar. Times such as 15.35 do not trigger the same automatic response and understanding as 3.35 pm.

As a traveller I am almost obsessive about times. I think one of my worse nightmares would be to arrive too late to catch my flight. So, the translation from 24 hour times to 12 hour times is one I take very carefully, which means that I double check (usually twice!)

The translation is, of course, two-way, 24 hour times to 12 hour times and 12 hour times to 24 hour times. We tend to meet the first translation more often.

Translating 24 hour times to 12 hour times

In a day the hour hand of a clock travels twice around the 12 hour face of the clock. There are 24 hours in a day.

The first 12 hours 59 minutes generate the same numbers for both the 12 hour and 24 hour times since the first circuit of the clock face is the same for 12 and 24 hour times. So 8 am is 08.00 hours,11.00 am is 11.00 hours and 12.45 pm is 12.45 hours.

At 1 pm, the 12 hour time returns to the numbers 1 to 12, but the 24 hour time is now into its second circuit of the clock and 1 pm becomes 1 plus 12 (from the first circuit), that is, 13.00. This continues, so 2 pm becomes 2 plus 12, that is, 14.00, 3 pm becomes 3 plus12, that is, 15.00 and so on.

Consider some comparisons / translations of key times;

12 hour time	24 hour time
12 midnight	00.00
6.00 am	06.00
12 noon	12.00
6.00 pm	18.00
10.00 pm	22.00

and other examples;

7.30 am	07.30
4.15 pm	16.15
10.30 pm	22.30

I am puzzled . . . Why do we say 'Twelve hundred hours' when there are sixty minutes in an hour, not one hundred?

The times where mistakes are most likely to occur are the times from 1 pm up to midnight. That is when the two systems become significantly different. So any 24 hour time between 13.00 and 24.00 could cause a problem. For example, a common mistake is to translate 20.00 as 10 pm (when it should be 8 pm).

To convert those 24 hour times which are in the 13.00 to midnight group in to 12 hour times, the first step is to subtract 2 from the 24 hour time to give you the key digit. (All the resulting 12 hour times will be pm).

For example, to translate 19.00 to 12 hour time, first subtract 2;

$$19.00 - 2 = 17.00$$

The 7 will be 7 pm when you have done the second subtraction, which is to take away 10 from 17.00;

$$17.00 - 10 = 7.00 \text{ pm}$$

As a double check, you could:

add back 12 to see if you return to 19.00 or

subtract 12 in one step, $19 - 12 = 7$

Translating 12 hour times to 24 hour times

Again, if you can picture a clock with the hour hand moving round for 24 hours, this should be reasonably straightforward. Up to 1.00 pm, the numbers are the same. From 12 noon, the 24 hour time is on its second circuit, but instead of using the 1 to 12 numbers again (in conjunction with pm) the numbers carry on from 12 to be 13, 14, 15, 16, 17, 18, 19, 20, 21, 22, 23 and 24.

To translate the times between 1.00 pm to 12 midnight, simply add 12.

Some examples;

2.40 pm becomes	$2.40 + 12 = 14.40$
7.15 pm becomes	$7.15 + 12 = 19.15$
10.30 pm becomes	$10.30 + 12 = 22.30$

16 Algebra

When you look down the list of maths topics that people find difficult and/or dislike then you quickly come to algebra. I think it gets bad press, undeservedly. Like many topics in maths you need to go back to the basics, the early maths, in order to understand algebra. It is not helpful to see algebra as something that just pops up and has to be dealt with as an entirely new topic.

When you look at algebra in text books you notice that there are letters as well as numbers. This may challenge security and consistency, but you do have to learn some new codes in order to understand algebra. Despite these codes and the letters, algebra continues to build on the basic rules you learned for working with numbers.

Algebra is really useful as a way of generalising and of communicating a generalisation. It is also used in many formulas. These are good reasons for developing some understanding of this topic.

Most formulas are examples of algebra

There are, then, some new conventions attached to algebra, some new codes to be aware of and learn. For example, *algebra does not use the × sign for multiply*. There are probably several reasons for this. I think that one of the reasons is that mathematicians like to present information as simply as possible in the sense of using a minimum of symbols to convey any information. It also helps to avoid a more obvious source of confusion. Algebra uses letters to represent quantities and one of the frequently used letters is x. Of course the symbol for 'multiply' or 'times' is × which is the same as the letter x. This is not true of the 'plus' symbol, +, which does not look the same as a letter The ever pragmatic mathematician abandons the × in algebra and a × b becomes ab and x times y is not written as x × y, but as xy. Similarly 'three times y is not written as 3 × y, but as 3y.

Neither is the division sign used in algebra. Division is communicated in the same format as it is in fractions, so x ÷ y is written as $\frac{x}{y}$ or $\frac{x}{y}$

An example of this 'not using × for times' is the formula for the area of a rectangle (any rectangle). To explain, start with the word version:

Area of a rectangle = length times width

Algebra often uses letters to represent a word/quantity.

In this case, if we use the letter A to represent area, the letter l to represent length and the letter w to represent width, the formula begins to look like algebra

$A = l \times w$

But the times symbol is not written in algebra, so the formula becomes

$A = lw$

This formula can be used to give the area of any rectangle. It is a generalisation of the way areas of all rectangles are calculated. This formula suits mathematicians since it is a very succinct way of expressing information. Mathematicians use the word 'elegant' as their top approval rating of a piece of mathematical work. It is most likely that one of the characteristics of an elegant solution is its brevity.

Algebra can be a much clearer way of expressing an idea. Look at the next sentence.

I described an idea in the times table chapter which showed that if you multiply two numbers together you get the same answer which ever order you use for the multiplication, for example 4 × 5 = 20 and 5 × 4 = 20.

A much shorter way of saying this in algebra is

$$xy = yx$$

x is used to represent one number and y represents the other number.

The equivalent idea in addition is

$$x + y = y + x$$

These are simple, minimal presentation equations which put over an idea in a clear way, providing you know the code. Of course, this is why all subjects have their own vocabulary. Once you know the vocabulary, the code, a complex idea is communicated clearly. If a physicist is told something is a transverse wave then she can immediately tell you a lot about that wave even before she knows its full identity. If you ask a guitarist to play an E chord, that is enough information for him to do that task. If you do not play the guitar, you are lost without a quick lesson from someone who does.

Since algebra uses letters to replace some numbers, the letters must behave like numbers and follow the same rules as numbers (and vice versa). This interchange of numbers and letters can help you understand and work out algebra.

For example, let's make a formula for the perimeter of a triangle.

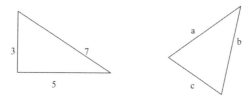

The perimeter means the distance round the figure, hence a perimeter fence. For the first triangle this is 3 + 5 + 7. For the second this is 4 + 7 + 9 and for the third the perimeter is, logically the sum of the lengths of the three sides, a + b + c. This now can be used as a formula for the perimeter of any triangle;

$$p = a + b + c$$

where a, b and c are the lengths of the sides and p is the perimeter. For an equilateral triangle, a triangle with three equal sides, each of length a, the perimeter, p is

$$p = a + a + a$$

which can be written as

$$p = 3a$$

This can be compared, but hopefully not confused with the formula for the volume, V, of a cube which has sides, each of length a

$$V = a \times a \times a$$

The algebra code does not use the times sign, so the equation becomes

$$V = aaa$$

which is written as

$$V = a^3$$

Now, let's make a formula for exchanging pounds into US dollars at a travel agent. This will introduce another item of algebra code, the bracket.

The exchange rate as I write this is $1.58 for £1. The travel agent will take a £3 commission fee for doing the exchange.

Start with a number example, say changing £103 into dollars.
Take off the commission £103 − £3 = £100
Now work out how many dollars 100 × 1.58 = $158

This was a two step process;
1) take away the commission
2) multiply the net amount in pounds by the exchange rate for pounds to dollars.

This can now be translated into algebra, using some code letters;

c   for commission
p   for amount of money in pounds
d   for amount of money in dollars
e   for exchange rate of £ to $
and brackets ( )

$$d = (p - c)e$$

This looks OK if you know the meaning of the brackets ( ), which are another one of the algebra codes. Some of the codes tell you an instruction, for example + tells you to add. Brackets tell you two linked instructions.

Brackets hold items together, almost like putting them in a box. The brackets around p–c, (p–c), tell you to treat this combination of p and c together.

Brackets also tell you about the priority of using the operations add, subtract, divide and multiply. In building the equation

$$d = (p - c)e$$

we subtracted (step 1) and then multiplied (step 2).

The rule / procedure of an algebra equation like this is 'DO THE BRACKET BIT FIRST.'

The brackets also mean multiply. In much the same way as the code lw means l multiplied by w, (p–c) e means p-c multiplied by e.

So (p – c) e tells you to

1) do the inside of the bracket first, that is the subtraction p – c
2) multiply the result of this subtraction by e

Which is exactly the procedure we used for the £103 to dollars example.

Algebra equations should always work when you use numbers in place of the letters.

This equation $d = (p - c)e$ would work for any money exchanged with a commission subtracted. You have to recode the letters, for example to change pounds into euros;

p   is amount in pounds (the same as before)
d   is the amount in euros
e   is the exchange rate of £ to euros.

The quadratic equation

This is a type of equation which is a top feature in early algebra. An example is;

$$A = (x + c)(y + b)$$

When you multiply the brackets out, the result is

$$A = xy + bx + cy + cb$$

I will try to show you how this is closely linked to the strategies we have used for times table facts and for long multiplication. This should demonstrate once again how the same ideas are recycled (in different disguises) time and again in mathematics.

Let's analyse the equation $A = (x + c)(y + b)$.

It is one number, $(x + c)$ times another number $(y + b)$. For example, remembering that letters represent numbers, x could be 20 and c could be 3, so $(x + c)$ could be 23. In the same way, y could be 40 and b could be 5, making $(y + b)$ as 45. So one example of a number equivalent to

$$(x + c)(y + b)$$

could be $(20 + 3)(40 + 5)$ that is 23×45

This is a multiplication which is the same as an expression for area ($A = wl$)

Now let's set up the area

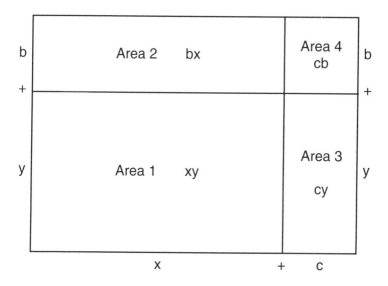

The area has four sub-areas, so the total area can be calculated by working out the area of the four sub-areas and then adding them together.

Area 1)   xy
Area 2)   bx
Area 3)   cy
Area 4)   cb

Total area, $A = xy + bx + cy + cb$

Try this again with another quadratic equation, this time a square.

$A = (x + a)(x + a)$

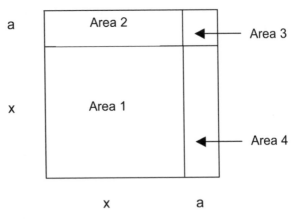

Area 1) x^2
Area 2) ax
Area 3) a^2
Area 4) ax

Total area, $A = x^2 + ax + ax + a^2$

The two ax terms combine to make 2ax, so the total area is

$A = x^2 + 2ax + a^2$

17 Famous Formulas

It can be very useful to be able to generalise some maths processes. That way you don't have to work from scratch every time.

A formula is a generalisation. For example, if I buy a bar of chocolate which costs 41p I spend 1×41p. If I buy two bars of this chocolate I spend 2×41p (82p). If I buy three bars, I spend 3×41p. This progression could become a little tedious, so I can generalise by developing a formula which I can use to work out how much I spend if I buy any number of chocolate bars which cost 41p each.

'The amount I spend' = '41p' × 'The number of chocolate bars I buy'

This can be written in an algebra format. If I use S to mean "The amount I spend" and n to mean "The number of chocolate bars I buy". The formula becomes;

$$S = 41 \times n$$

You now have to remember another piece of the code (as mentioned in the previous chapter). The convention (normal practice) in maths formulas which use letters to represent the factors involved (in this case the letters used are S and n) is to not write the times sign (×). This works on the principle that if it's not (written) there you assume it is there!

The formula becomes: $S = 41n$

Circles

If there is a scene in a film or play set in a classroom or lecture theatre, especially if it is a science or maths class, then there will be a blackboard and written on it will be $A = \pi r^2$. This is the formula for the area of a circle, using r to represent the radius.

π (pi) is a special number, not a whole number but a number which is a bit more than 3 and can be written to so many decimal places that this activity can get the successful (and persevering) person (though really it should be their computer) into the Guiness Book of Records.

This special number, which has its own symbol, π, is called pi. Pi is to do with circles. It is a sort of Glastonbury number . . . Natural, Man.

If you want to try and find an approximate value for pi, take a piece of string and wrap it once around a biggish tin can. This length of the string is called the circumference. Measure it. Now, as accurately as you can, measure the diameter of the can

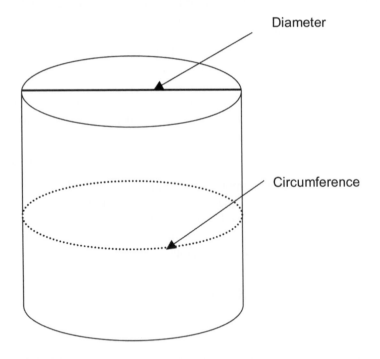

Diameter

Circumference

Divide the circumference by the diameter and the result should be an approximate value for pi. A value, which is acceptably accurate for most calculations, is 3.147.

Pi is the ratio of the circumference of a circle to its diameter, that is, the circumference of a circle divided by its diameter.

$\pi = C / d$

It is the same for ALL circles.

You can turn this formula around to

$C = \pi d$

so you can calculate the circumference of any circle if you know its diameter.

Pi also features in the formula for the area of a circle, as I said in the opening paragraph of this chapter.

Area of a circle $= \pi r^2$

For some people the words *formula* and *algebra* create an immediate barrier. I think this is another example of the vocabulary and symbols of maths adding a mystique which keeps some learners away. This use of a special vocabulary and symbols is not unique to maths. You can find it in many places, for example, computer users speak of RAM and KB, language specialists talk about graphemes and phonemes, car experts talk about torque. If you have the inclination and if someone explains these words, without using even more specialist words in their explanation, then you can understand.

Pythagoras

The Pythagoras equation (or formula) is sufficiently famous for it to be the subject of a rather long and somewhat contrived joke, which I am not telling here, involving a squaw and a hippopotamus.

Pythagoras was a Greek who lived about 2500 years ago. He set up a secret religious society which explored the mysteries of number. He believed that the study of arithmetic was the way to perfection (something governments seem to believe when discussing numeracy).

Pythagoras was not the first to discover the theorem which now bears his name. The Chinese used it for surveying and the

Egyptians used it to help build the pyramids. This famous theory has enabled builders to produce set squares of exactly 90°. It has to be a pretty useful equation. Thus there is a direct link between the Pythagoras equation and a 90° angle.

For a nation of such great architects as the ancient Greeks, an accurate measure of 90° would have been essential. A consequence of the Pythagoras theory was a special case of a right angled triangle, the 3, 4, 5 triangle. This is a triangle whose three sides are 3 units, 4 units and 5 units. If these sides are measured accurately and joined accurately, the resulting triangle includes an exact right angle.

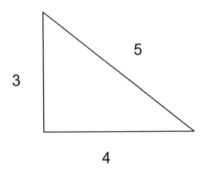

Pythagoras's equation deals with the lengths of the three sides of any right angle triangle. If you square the lengths of the two smaller sides and add those two squares together then that gives the same number as the square of the length of the longest side. (This longest side, opposite the right angle, is called the hypotenuse).

$$a^2 + b^2 = c^2$$

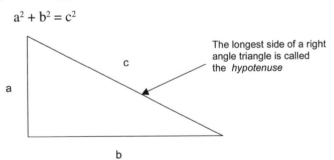

In the 3, 4, 5 triangle, the shorter sides are 3 and 4.

The square of 3 is 3 × 3, that is 9.
The square of 4 is 4 × 4, that is 16.

The sum of these two squares is 9 + 16 = 25.

The square of the longest side is the square of 5 is 5 × 5, that is 25.

Thus, "the square of the hypotenuse (longest side) is equal to the sum of the squares of the other two sides."

For the practical construction of a right angle, this 3, 4, 5 triangle is the simplest example of Pythagoras's equation. It is possible to have multiples of 3, 4 and 5 such as 6, 8 and 10 (where all three sides have been doubled).

The next whole number example of this special equation is a little less practical in the sizes of its sides. The three sides are 5, 12 and 13.

The two short sides are 5 and 12.
The square of 5 is 5 × 5, that is 25.
The square of 12 is 12 × 12, that is 144.
The sum of squares is 25 + 144, that is 169.
The longest side is 13.

The square of the longest side (the hypotenuse) is 13 × 13, that is 169.

It is fascinating to think that these calculations are thousands of years old. Number facts endure!

18 People Having Difficulties With Maths

This chapter was written by Pete Jarrett who is a Curriculum Support Tutor and Teacher at Truro College in Cornwall. He specialises in helping learners in Further and Higher Education who have difficulties in Mathematics and the Sciences.

Case Studies

The stories of people with dyscalculia and other maths learning difficulties are of great value. Numeracy is an aptitude and skill society often takes for granted. Case studies can enable those who are numerate get a glimpse into another world. For those who wrestle daily with numbers, hearing stories that echo their experience can be comforting. They help to make people feel that the everyday struggles they endure are not unique, strange, or particularly different. The value of sharing stories should never be underestimated, especially in helping deal with the anxiety and stress a maths learning difficulty can bring.

These case studies aim to represent the person's voice, supported by my comments from a professional perspective. Four stories, each different, yet with many similarities. Whilst they cannot cover every aspect of dealing with dyscalculia, they are intended to highlight some of the common experiences people encounter.

The first story is that of Elisabeth. Although Elisabeth describes a significant difficulty with mathematics, she has only recently identified her "I don't do numbers" as a learning difficulty. So we have used her story as a way to look at some of the areas that an assessment might cover in terms of identifying a difference. We then move to Bethany's story to hear her experiences in school and how she has managed to deal with her difficulties in everyday life. Sally's story looks at some of the teaching techniques used to get her through an important exam. The final story is that of Alexa, and I hope that this story of compassion, determination and collaboration acts as a starting point for a coordinated approach to dealing with dyscalculia in the future. So, let's start with Elisabeth's story.

Elisabeth

"I often wish I could be taught maths as if I were two again. And only progress once I had fully grasped the basics."

I have included Elisabeth's story because she has never had a diagnosis, or been through any screening process for a maths learning difficulty. Her story provides an opportunity to discuss what might be involved in identifying a maths learning difficulty in a non-educational environment.

Elisabeth is a journalist published in all the UK broadsheets and most women's magazines, and is now a digital editor - her blog was shortlisted in the 2009 Guild of Food Writers Awards. Like many people with maths learning difficulties, her communication skills and creativity are highly developed. This discrepancy between abilities in other disciplines and an inability in numeracy or mathematics is one of the primary indicators of a maths learning difficulty. The existence, or not, of a discrepancy would be one of the first points to establish in any screening discussion. Discrepancies in performance are often apparent irrespective of an individual's wider intelligence, and can be relatively easily established. They also constitute an important element of an education psychologist's diagnosis, which is able to take a more objective and thorough assessment of the cognitive performance or various intelligences of an individual.

In my first conversation with Elisabeth, I was keen to establish how she saw her problems with maths. I wanted to know 'what happens when you try to do maths?' and 'what psychological or emotional impact has your problems with maths caused?'

Traditional maths assessments tend to look at what you can do, but we are also interested in what you can't do. For example, Elisabeth says "I never got that system of rounding-up to 10 to make mental arithmetic easier. I still count on my fingers". Rounding-up numbers to make estimating easier is a functional numeracy skill that many people take for granted. It also demonstrates an ability to use the intuitive number line that appears to be absent in developmental dyscalculia.

Neuroscientists who are studying the development of arithmetical and numerical cognition, such as Professor Stanislas Dehaene in Paris and Professor Brian Butterworth in London, have pointed towards a sense of number, and a seemingly intuitive representation of what they call numerosity that seems either to be absent or to work in a different way in people with developmental dyscalculia. When I asked if Elisabeth could offer any insights in to how she understood number, she described how, for her: "Numbers up to 10 have personalities (8 is warm but 9 is remote), a number such as 7 come up as a large S in my mind; 29 and 32 fox me because the 9 makes the 29 look like bigger than 32. I can however do my 2, 5 and 10 times tables, not that this has helped me with any other calculations."

She goes on to say; "I might attempt to add 7 + 9 in my head but I can't guarantee I will get the answer – I still remember the look of surprise on one colleague's face when I used my fingers." This might be a good time to talk about the use of fingers – we use a decimal number system, the patterns work in multiples of ten, we have ten fingers. It seems entirely practical to use these digits as provided. The only reason we don't use them is that society dictates that it is wrong. If you keep using fingers you may well find that you begin to build a visual representation in your head which means you can do the sum without taking your mittens off – but other than that, don't worry about it! Keep using your fingers!

Generally, the struggles Elisabeth describes are sufficient to suggest the likelihood of a maths learning difficulty (I don't always like using the term dyscalculia because other conditions can affect the learning of mathematics). In our discussions I asked her a number of further questions in the form of the Checklist which is included in this book; she answered many in terms of the strength of the word yes, rather than yes or no.

Elisabeth did not take maths at O-Level: "It was taken for granted there was no point. I did biology instead and loved it. It's interesting to think now how having dyscalculia may have influenced my life choices."

We discussed the emotional aspect of her difficulties; "even looking at the word 'maths' makes me wince. This word is not for me. I turn my head away. I have other ways of feeling enriched: through creativity and words. "

She also sees a benefit in having a problem with numbers: "I think it has made me a good editor: I don't assume people know what I know; I am not complacent about my accuracy - I obsessively fact-check. I also write as clearly as possible because I know the obvious is not always easy to understand."

She learned to accept her inability as a fact-of-life ("Thank God for calculators" she says) but the recent assessment has been a boon. "It feels like a breakthrough. I feel vindicated - I am not stupid or vague. I used to think I had a psychological block. Knowing it is a learning difficulty makes numbers easier to deal with, because I feel more confident."

I don't always advise continuing beyond a screening to a full diagnosis as it doesn't work for everyone, but in Elisabeth's case it might be the logical way forward. She would certainly welcome it: "I'd love a label. It would help me explain myself to others, too."

BETHANY

Bethany was in her final year at secondary school when her SENCO contacted me to ask about screening for dyscalculia and maths Learning Difficulties to access exam concessions. It was very apparent that Bethany's performance in maths was far below that of her performance in other subjects. Her Head of Mathematics noted "her methods, even solving everyday problems, are often particular to herself, following a different path from most students". This is exemplified in her visual interpretation of her method of division.

What was clear was Bethany's determination to gain her goal of a grade C at GCSE, her Head of Maths saying "Beth will often need more time to complete a problem and has been allowed this time during lessons throughout her time at this school. She will often still be working at the end of a lesson and work through, when other students are leaving."

Exam concessions, also known as access arrangements, are available to students in order to lessen or remove the effects of a 'substantial disadvantage' in an assessment. For example a Braille paper is an access arrangement which would be a reasonable adjustment for a Braille reader, but not for a candidate/learner who was unable to read Braille. Access arrangements allow candidates/learners to show what they know and can do without changing the demands of the assessment. Generally, in the case of learning difficulties relating to mathematics extra time can be available to students who may need time to allow for a slower processing speed or because they need to employ strategies that may require time to put in place – such as Bethany's method of division. To obtain concessions a number of factors need to be identified. We have already seen that Bethany was allowed time in school to complete problems, and that this was her normal way of working, therefore, if she is allowed extra time in her studies at school, it is recognised that her processing speed is slower than other students performing at the same level, and so it is fair that this is recognised in an exam. It is also helpful to assess a learner's performance against a standardised scale of cognitive performance. There are a number of assessments that are used including the WRAT4 and the British Ability Scale II. In essence they all perform the same function, and that is to identify a learner's performance in relation to other learners in her age group.

Bethany was assessed using the British Ability Scale II. Her verbal reasoning skills were at the 98th percentile, representing an age equivalent score of over 18 years, well above the average range expected for a pupil of her age. However, on the Number Skills scale she scored on the 28th percentile, which is towards the bottom of the average range expected for a student of her age, and with an age equivalence of 11 years and 3 months. Apart from the very obvious discrepancy between Bethany's numeracy skills and her verbal reasoning skills, the Educational Psychologist also made reference to the high anxiety levels and panic attacks that Bethany had suffered around mathematics.

Based on the evidence of her normal way of working from the school's Head of Mathematics and the report from the Educational Psychologist Bethany was awarded extra time in her maths exam, and, in what she describes as "a minor miracle" she gained a 'C' grade.

It is clear that Bethany's difficulties with maths have had a profound effect on her. She has suffered from severe anxiety for many years, including panic attacks, especially if she has to do some form of mental arithmetic test. It is when you talk to Bethany about everyday tasks that you begin to realise how different her view of the world is from what most of us experience. In Bethany's words; "Money is a big problem. When I am paying for something in small change and haven't worked out what coins I need before it can take me ages to work it out (much to the amusement, or more often irritation, of the person behind the till!). Under the stress of the situation I often forget what change is already in my hand and have to put everything back into my purse and start again. Most of the time I try to work out amounts before I pay or just carry a lot of pound coins, as they are easier to count. I quite like 99p signs. When I see one of these I know it means I can take out a pound coin and everything will be fine. At least with 99p I can't mix up the numbers because they are the same."

Bethany also has great difficulty with reading bus timetables, finding the information around the time she needs to be very distracting. "I generally just go to the bus stop and hope one will turn up – which can mean leaving a lot earlier than necessary if I have to be somewhere on time".

I asked Bethany if people had been accepting of her difficulties. "I think that people tend to accept some things more than others. If I said that I struggled with timetables, for example, people might be a bit more accepting of this than if I said I couldn't read the time. People tend to find this very strange, and sometimes if I slip up and say 45 past instead of quarter to, they also don't understand that".

Like many people with a maths Learning Difficulty Bethany finds it hard to put into words what she experiences, partly because some people don't understand, but also because "describing my problems can sometimes be hard, as to do that I need examples, which I often can't provide as the examples are the exact things I find hard in the first place. Sometimes I think it would be nice to borrow somebody else's head for a bit, just so I can explain".

Finally, I asked Bethany if she has a piece of advice to share with other people with a maths learning difficulty.

"Probably that sometimes there are ways around certain maths things, it's just sometimes hard to find the way".

Sally

Sally was undertaking a PGCE qualification when I met her. One of the requirements of the qualification was that students had to achieve a Level 2 qualification (An equivalent standard to a GCSE pass at grade C) in both Literacy and Numeracy. Sally was concerned about her ability to pass the numeracy exam. We agreed to meet for a one to one session to discuss her concerns, and to enable me to get an insight into her readiness to take the test. Sally said that maths caused her great anxiety and she was very concerned that her whole career path was at risk because of the maths test.

I asked Sally to look at some practice questions similar to those she would encounter in the actual test. The first couple of questions went OK, and whilst Sally was much slower than I would normally expect, I didn't suspect much more than severe anxiety. It should be stressed that maths anxiety is a very real and significant problem that needs careful consideration and support, and the impact of anxiety on a learner with maths Learning Difficulties, or indeed any learner who finds maths difficult should never be misunderstood.

However, the third question required the learner to extract information from a table with four columns and three rows. Sally was suddenly pitched into a world that made her feel exceedingly uncomfortable. She was unable to follow down columns and along rows to get to the values we needed. In situations like this it often helps to use rulers or pieces of paper to help underline the required data, but Sally found it impossible to even link the column and row headers to allow this strategy. Based on the preliminary discussions and these problems it was clear that Sally would benefit from an assessment from a psychologist, and so I referred her immediately. Sally wasn't that keen to be assessed as she felt it confirmed the differentness that she had always felt, and had never been comfortable with. She continues to be unsure of the benefit of a diagnosis as she doesn't feel it has changed much. The one slight silver lining was that she was entitled to extra time concessions for her upcoming test, she wanted to pass that because it had to be done, but after that she wanted to continue living a relatively maths free life. In a previous career she had to employ some fairly complicated strategies to allow her to do her job safely, which had placed a significant amount of stress upon her – she and mathematics had their time together, they had to meet again briefly, and hopefully after this test they could finally go their separate ways.

Perhaps some people will be surprised to hear that people with significant maths learning difficulties can pass maths exams. Some can, some find it incredibly difficult and that the qualification is of little value, and therefore don't see the difficulty as an acceptable trade off for the gain. Sometimes a qualification needs to be achieved to allow for a career choice to be made. In these circumstances I never cease to be inspired by the work that learners put in, and the complexity of the strategies that they employ to overcome their difficulties.

Like many learners with a difficulty Sally had very well formed problem solving skills, she also had the advantage of having a good level of literacy that would provide a useful tool. In fact, it is not always the complexity of a mathematical problem that will cause difficulties, but the very basic numeracy skills. Because the value or magnitude of numbers was hard for Sally to follow, with her regularly losing track of calculations involving more than ten's and units, we used place value columns for all calculations. For the exam, the first thing Sally did was to draw out a series of columns that she would be able to use for calculations.

Sally had no concept of number bonds, that $7 + 3$ and $3 + 7$ will both always equal 10, or that $17 + 3$ will therefore equal 20. Every time she saw a sum like this she was seeing it for the first time. The same was true of the times tables. The sheer volume of number facts that have to be remembered can seem overwhelming, or even impossible to learn. We therefore spent a considerable amount of time working on a reduced number of number facts and rules that would hopefully simplify calculations. Our use of place value columns helped us to see that each column, units, tens, hundreds and beyond obeyed the same rules – we regularly used number lines to reinforce this, and we went right back to basics, learning by rote the number bonds to 20. Whilst rote learning may have gone out of fashion in some circles it can be a particularly effective, but sometimes slightly boring, tool. We also reduced the number of times tables we used, relying on the 1, 2, 5 and 10's, and breaking down multiplications into manageable chunks 12×17 became 10×17 (add a zero = 170), 2×10 (20) and 2×7 (14), with the answers added together (204). The place value columns helped to keep control of the sums. This was a time consuming process, but of course this is what the extra time was for.

Alongside these strategies we concentrated on the areas that Sally found easier to understand, and perhaps neglected some of the harder areas. In what was a calculated risk we worked out what we expected the pass mark to be, and what were the most likely types of questions to come up, playing to our strengths. We were looking to pass. We only had to hit the pass mark; we knew we could afford to leave some questions, a few at least, if they were causing stress of taking an unreasonable amount of time. Sally stuck to this strategy – and she passed! She is now an exceptional teacher in an FE College and is a great advocate for any student with a Learning Disability.

Alexa

Alexa has a diagnosis of dyscalculia and Non-verbal Learning Disability (NLD) from an educational psychologist. Her diagnosis of dyscalculia is supported by a confirmation from University College London. Like many dyscalculics, Alexa is extremely articulate, and campaigns to promote awareness of dyscalculia with passion. She is a projectionist and artist, and much of her work focuses on how her learning difficulties have allowed "a different way in processing the world". Perhaps it would be best to allow Alexa to describe her differences in her own words.

"I knew there was something different from quite a young age, I do not remember when exactly but my difficulties in maths were the first thing to surface. I remember feeling physically sick going into my maths lessons in junior school. Throughout this time, and later in secondary school I was stuck in this perpetual loop where people would expect a lot from me, I would tell them I was struggling, they would tell me to try harder, I would try immensely hard but would still not meet the expected outcome.

From what I have come to understand, dyscalculia affects my ability to subitize, (subitizing relates to what seems to be an innate ability to recognise the magnitude or value of a number. Most of us intuitively recognise up to about four objects without counting, and this represents a sense of number that is apparently absent in people with developmental dyscalculia) and to gauge amounts, distance and time. It is the distinct lack of number sense. I think not being able to tell left from right sits here as well. My NLD also affects the comprehension of mathematical word problems, such as 'Sally went to the shops to buy six apples with a five pound note. . . .' I often read things like this and go 'huh?'". Alexa asked me if I could come up with a better mathematical problem, but I have left it as she wrote it. Many people struggle to make sense of questions in which the maths is hidden, and this is a common issue with people with NLD and Aspergers Syndrome, and also dyslexics. It is another aspect of maths learning that is poorly often unrecognised or appreciated. Of course this is an issue that is not unique to people with maths learning difficulties, many students find it hard to extract the maths from word problems, the mix of maths and English is the joining of two entirely different languages into a new, often unintelligible language of Numglish.

Alexa is very open about the upset and stress caused because her struggles were not recognised until she was an adult. "I think the emotional impact has been quite substantial; I have suffered from depression, anxiety and low self-esteem. For a long time I honestly thought I was 'stupid'. I have come a long way in overcoming these feelings but they do resurface on bad days, usually when someone has been ignorant or impatient regarding something I was taking time over, such as counting out change when shopping. If it had been caught early; all of the stress and emotional problems I have had to deal with could have been prevented from a young age. It troubles me to an even greater extent that there are countless people in the same position now, some uninformed of dyscalculia, struggling due to unawareness."

Alexa is very active in the Dyscalculia Forum (www. **dyscalculiaforum**.com), recognising the incredible value of shared experiences; most of the people on the Forum have, or think they may have, dyscalculia or a maths Learning Difference, although there are teachers, educators and academics who use and contribute to the forum as well. Alexa describes her involvement: "the first type of support I received from the forum was information. I was quite shy at first and spent most of my time reading through back dated posts. I remember being struck at how similar my struggles were to other peoples all over the world. When I had the courage to introduce myself I found the forum to be a nurturing and supportive space. After becoming public about my struggles I started to receive a lot of support and messages of thanks from other people who sympathised and recognised that my struggles were the same as their own. The fact that writing about my experiences could help others in this way pushed me to become more and more open about my struggles".

As a professional working with people with maths Learning Difficulties it is sometimes easy to forget that people with differences need to be involved in the process of understanding and defining difficulties with mathematics. Alexa has invested a significant amount of time and money trying to find professionals who were and are able to help her to interpret her view of the world, and in doing so has become as well informed, if not better informed, than some of the people that make judgements on maths learning difficulties. One of her greatest frustrations relates to the sometimes confusing, and differing opinions of professionals. "Unfortunately, until someone takes a global stance as to what dyscalculia and its types, or maths learning disorders, or difficulties, actually encompasses then we are going to be wading through a lot of different terminology. This is where well informed support can get a bit hazy and it has taken me quite a lot of searching and reading to wade through it all". Accessing the right support will not always be easy until there is universal recognition of the need to manage the impact of maths Learning Difficulties, and a greater understanding of how significantly such struggles can affect the lives of those with such problems.

If the case studies in this book have achieved anything I would hope that they have demonstrated that many people are able to find very effective strategies to cope in most aspects of their life, but, often at a huge emotional cost, and often in relative isolation. Much can be gained from sharing experiences, and these all add to the understanding of life with dyscalculia, which will in turn help to increase awareness and ultimately knowledge of such difficulties. We are only at the beginning of a journey, we are gaining knowledge quickly, but we still have a way to go, and we must continue to make this journey together.

19 The Dyscalculia Checklist

The Checklist is about identifying behaviours, skills and deficits. It can act as a basis for setting up a programme of intervention and help. There is not a definitive 'score' that would classify a learner as dyscalculic. The list is more subjective than that.

© Steve Chinn, 2008

Name _____ dob _____ date _____

Does the learner . . .

1) Have difficulty counting objects accurately. ☐

2) Lack the ability to make 'one to one correspondence' when counting (match the number to the object) ☐

3) Find it impossible to 'see' that four objects are 4 without counting (or 3, if a young child) ☐

4) Write 51 for fifteen or 61 for sixteen (and all teen numbers) ☐

5) Have difficulty remembering addition facts ☐

6) Count on for addition facts, as for 7 + 3, counting on 8,9, 10 to get your answer ☐

7) Count all the numbers when adding, as for 7 + 3 again, you count 1,2,3,4,5,6,7,8,9,10 ☐

8) Not 'see' immediately that 7 + 5 is the same as 5 + 7 or that 7 × 3 is the same as 3 × 7. ☐

9) Use tally marks for addition or subtraction problems ☐

10) Find it difficult to progress from using materials (fingers, blocks, tallies) to using only numbers

11) Find it much harder to count backwards compared to forwards

12) Find it difficult to count fluently less familiar sequences, such as: 1,3,5,7,9,11 . . . or 14,24,34,44,54,64 . . .

13) Only know the 2×, 5× and 10 multiplication facts.

14) Count on to access the 2× and 5× facts

15) Learns the other basic multiplication facts, but then forget them overnight

16) Make 'big' errors for multiplication facts, such as $6 \times 7 = 67$ or $6 \times 7 = 13$

17) Find it difficult to write numbers which have zeros within them, such as 'four thousand and twenty one'

18) Find it difficult to judge whether an answer is right, or nearly right

19) Find estimating impossible.

20) Struggle with mental arithmetic.

21) 'See' numbers literally and not inter-related, for example, do you count from 1 to get 9, rather than subtracting 1 away from 10.

22) Forget the question asked in mental arithmetic

23) Prefer to use formulas (when you remember them!), but use them mechanically without any understanding of how they work

24) Forget mathematical procedures, especially as they become more complex, such as decomposing or borrowing for subtraction and almost certainly any method for division.

25) Think that algebra is impossible to understand

26) Organise written work poorly, for example does not line up columns of numbers properly.

27) Have poor skills with money, for example, are you unable to calculate change from a purchase. ☐

28) 'Think an item priced at £4.99 is '£4 and a bit' rather than almost £5 ☐

29) Not see and pick up patterns or generalisations, especially ones that are new to you, for example that ½, ⅓, ¼, ⅕ is a sequence that is getting smaller. ☐

30) Get very anxious about doing ANY maths ☐

31) Refuse to try any maths, especially unfamiliar topics. ☐

32) Become impulsive when doing maths, rather than being analytical. Do you rush to get it over with? ☐

Appendix

Two words from the maths vocabulary used in this book

I have used the words 'digit' and 'number' quite specifically to mean;

Digit. Any of the symbols, 0, 1, 2, 3, 4, 5, 6, 7, 8, 9

Number. Any of the digits, used individually or combined to represent a value. For example, 14 is fourteen, 506 is five hundred and six. (see also place value).

Other maths books by Steve Chinn

"Mathematics for Dyslexics: A Teaching Handbook" 3rd edition (2007) Chinn, SJ and Ashcroft, JR, published by Wiley ISBN 0-470-02692-8

"The Trouble with Maths" (2004) Steve Chinn, published by RoutledgeFalmer. ISBN 0-415-32498-X (Winner of the NASEN/Times Educational Supplement 'Book for Learning and Teaching Award' 2004) Second edition to be published in 2011.

"More Trouble with Maths: Assessing Maths Learning difficulties and Dyscalculia" will be published by RoutledgeFalmer at the end of 2011

"What to do when you can't learn the times tables" Steve Chinn, published by Egon

"What to do when you can't add and subtract" Steve Chinn, published by Egon

www.stevechinn.co.uk